I0568943

# COOL GREEN LEAVES & RED HOT PEPPERS

*Growing and cooking for taste*

CHRISTINE McFADDEN
MICHAEL MICHAUD

*Photographs by* JAMES MERRELL



FRANCES LINCOLN

*This book is dedicated to Joy Michaud, whose gardening skills made it possible.*

Frances Lincoln Ltd
4 Torriano Mews
Torriano Avenue
London NW5 2RZ

**Cool Green Leaves & Red Hot Peppers**


First Frances Lincoln edition: 1998



British Library Cataloguing in Publication data
A catalogue record for this book is available from the British Library.

ISBN 0 7112 1223 6

Set in 11/15pt ITC Legacy Serif Book
Printed in Hong Kong

2 4 6 8 9 7 5 3 1

**Cook's Notes**

The metric/imperial conversions used in the recipes are precise.
The two sets of weights are therefore interchangeable.
Spoon measures are level.
Herbs are fresh, unless dried are specified.
Vegetables are medium-sized, unless specified otherwise.
Eggs are large

# INTRODUCTION

The last few years have seen an amazing change in our attitude to vegetables. The popularity of Mediterranean, Middle Eastern and Pacific Rim cuisines, the awareness of the links between diet and health, the growing sophistication of vegetarian cooking, as well as travel to exotic destinations, have all played a part. Vegetables are now living up to the meaning encapsulated in the Latin origin of their name – *vegere*, meaning to grow, animate or enliven.

Cooking techniques are changing too. Briefer, lighter methods such as stir-frying, chargrilling and steaming mean that we can enjoy the life in vegetables to the full, making the most of all the delicious flavours, textures and colours that vegetables offer us. The range of varieties – both unusual and familiar – now available from seed merchants, greengrocers and supermarkets is becoming more and more exciting, opening up wonderful opportunities to both the gardener and the cook. Specimens once thought too fragile to grow anywhere outside the tropics have been found to flourish quite happily in chillier northern climes, given the right conditions. We have successfully grown chillies, aubergines and peppers, as well as lemon grass and Mexican tomatillos.

Thanks to improved transportation and storage, many unusual and wonderful vegetables from the Middle East and Asia are becoming a familiar sight in western supermarkets, and if you buy from shops catering for the ethnic communities the possibilities are even greater. Bottle gourds and bitter melons, yard-long beans and white aubergines, black radishes and mustard greens are all there to tempt us, along with a wonderful variety of herbs.

This vegetable cornucopia provides us with an overwhelming choice, particularly when growing from seed. In order to make informed choices, the gardener needs to know which variety will grow best in a particular soil or climate, and the cook needs to know which variety is best for the job intended. For example, most cooks know that floury potatoes are good for roasting and waxy ones are best for salads, but when it comes to deciding which type of tomato has a firm enough flesh for stuffing, or which would be the perfect pepper for a stir-fry, we're on less familiar ground. We have spent many enjoyable hours experimenting in the kitchen and assessing how different varieties of the same vegetable behave. Most of the vegetables used for testing we grew ourselves, so we are speaking from first-hand experience.

Our passion for vegetables stems equally from a love of growing them, a desire for produce of the best possible quality and a deep-seated belief that what you eat and cook reflects the way you feel about yourself and the people important to you. We hope, therefore, that the recipes in this book will inspire you to experiment both in the garden and in the kitchen. We also hope that we have in some small way helped bring about the late Jane Grigson's vision of 'high-rise blocks, patched with vegetation on every balcony . . . every town ringed again with small gardens, nurseries, allotments, greenhouses, orchards, as it was in the past, an assertion of delight and human scale' *(Jane Grigson's Vegetable Book*, 1978*)*.

## Horticultural matters

This book is not, and was never intended to be, a conventional gardening book. We have simply given general guidelines, with, where appropriate, special cultivation and harvesting tips, based mostly on our own experience. Standard horticultural information can be found elsewhere. There are several good books available – one of the best is Joy Larkcom's *The Salad Garden* (Frances Lincoln, 1984), which has been a constant source of inspiration.

Nor does our book include every vegetable. We started from the basis that we should be able to grow the vegetables ourselves. Thereafter, a certain amount of idiosyncrasy came into play. For example, we left out asparagus simply because we don't find it inspiring, but we included ocas and exotic cucurbits because they fascinate us. Some varieties we have been growing for years and these were naturally included; others were recommended by gardening friends and magazines. Since we are always on the lookout for something new and unusual, we also depend on the seed catalogues to some extent. Ultimately, though, there is a certain amount of luck, both good and bad, in sourcing good vegetables. This is, of course, part of the fun and challenge of gardening, and we will continue to make new discoveries.

On the question of pesticides and fertilizers, we prefer to grow and buy vegetables that have been exposed to a minimum amount, organic or otherwise. Given due diligence and healthy soil that is in good heart, high-quality vegetables can be grown with only limited and judicious use of chemicals.

### Plant classification

The plant-grouping system used by professional botanists is based on the similarity of certain characteristics, such as flowers and fruit. This branch of botany is called taxonomy, and is useful in clarifying the confusion that would otherwise occur.

One of the main groupings used by taxonomists is the 'family'. An example is the Allium family, which includes onions, leeks and garlic. Peas and beans, on the other hand, belong to the Legume family, and sweetcorn belongs to the Grass family. Families are further divided into smaller groups called 'genera', with each genus sharing certain traits that distinguish it from other genera. Genera are then divided into 'species', each of which has its own special characteristics.

Species are broken down into further divisions: that with which gardeners are most concerned is the 'cultivar'. This refers to a distinctive group of plants that share traits that may be desirable to the gardener. The term is a contraction of 'cultivated variety' and replaces the old-fashioned 'variety'. We feel 'variety' is a term more familiar to readers and have therefore opted to use that rather than 'cultivar'.

### Sourcing seeds

The majority of seeds of the varieties listed in this book are available from standard seed catalogues. Some come from specialist suppliers, while a few have been sourced from companies in other countries. To find the best varieties gardeners must be prepared to actively seek out what they want and, if necessary, to approach unfamiliar seed companies directly with their requests. Two invaluable reference books that list varieties as well as seed companies are *The Seed Search* by Karen Platt and the *Garden Seed Inventory* compiled by Kent Whealy of the Seed Savers' Exchange (see Bibliography).

### Tasting trials

In order to create a true picture of a vegetable's properties – flavour, smell, texture, and so on – and to assess the effect of the growing environment on eating quality, tasting trials would have to be carried out over a period of several years on varieties grown at different sites, using teams of trained tasters. It was impractical for us to conduct such meticulous tests, and our results should be looked on as subjective guidelines only.

Where vegetables needed cooking, we used the simplest methods possible, such as steaming, blanching or frying in a neutral oil, so as not to detract from the inherent sensory properties of the vegetable.

As a result of our taste tests, we became aware of the limitations of language. For example, the terms 'pungent', 'peppery' or 'chewy' may have different meanings attached to them by different people. In the world of vegetable tasting there is, as yet, no official language equivalent to the descriptive terms used in the tasting of wine or olive oil. The time is ripe for the development of universally accepted tasting terms.

### Gardening in a limited space

The most obvious reason for growing your own vegetables is to enjoy harvesting and eating absolutely fresh produce. Not only do vegetables pass directly from plot to pot, but you also have the chance to try varieties that are not normally available in the shops.

In an ideal world, your vegetable patch is best located close to the house. However, the lawn may be too precious to dig up or the garden may be no more than a concrete patio. In Britain, an allotment might be a possibility. Despite their noble history (they did, after all, produce significant quantities of food for the British during both world wars), allotments have an ill-founded reputation for stodginess; in fact, sites can be lively and vibrant places, representing the diversity of the local community. If you are willing to give an allotment a try, you should contact the local authority to check on the availability of a plot. Rents are modest, and a plot is usually big enough to supply at least some vegetables nearly all year round.

If the only area you have is a balcony or patio, then try growing vegetables in containers. These come in various shapes, sizes and materials, from purpose-built pots of terracotta, wood or plastic, readily available from garden centres, to the huge plastic tubs supplied by specialists in horticultural sundries. For the more ecologically minded, throw-away items such as bath tubs, sinks and barrels can be recycled as miniature vegetable gardens.

Containers can be planted up according to culinary themes. Try a tub of Mexican plants – chillies, cinnamon basil, tomatillos and epazote – or a Thai tub of lemon grass, coriander and holy basil. That way, whenever you want to cook an ethnic meal, you'll have everything you need growing in one place.

### Ornamentals

Vegetables with particularly attractive features can be successfully integrated into the border as ornamentals, satisfying both the painter's palette and the cook's palate. Puns aside, plants with both edible and ornamental value include 'Rhubarb Chard', the dark green leaves of which are contrasted against bright red stems, and the 'Painted Lady' runner bean, which has bicolour flowers of red and white. Properly planned and planted, these edible borders can be visually pleasing while at the same time providing quality vegetables for the most discerning of cooks.

### Protected cropping

Protecting vegetables from the vagaries of the weather can bring about an earlier harvest and, in some varieties, increase yields. Flavour may also be noticeably improved.

One of the cheaper options is agricultural fleece, a thinly spun synthetic fabric that can be laid over a crop, provided plants are fairly low-growing. It acts rather like a blanket, holding the heat around the plants while protecting them from chilling winds. Because it is a fabric, the fleece can simply be folded for easy storage when not in use.

Cloches are another alternative for low-growing crops. Some are made of flexible polythene stretched over hoops embedded in the ground, while others are moulded plastic or panes of glass or stiff sheets of plastic attached to a frame. Cloches can be removed when temperatures begin to rise.

For those with the space, the most effective – though the most expensive – way to grow protected crops is in a greenhouse or a polytunnel. The greater height will accommodate taller vegetables, such as tomatoes and climbing French beans, and if necessary plants can be grown in containers.

## Vegetables and health

Though nutrition is a constantly evolving science and fads seem to come and go overnight, scientists are now providing proof of what the ancient healers already knew – that vegetables are vital to good health. Modern nutritional research reveals, time and again, a strong and consistent link between a diet high in vegetables and a decreased risk of many life-threatening diseases, particularly cancer. Vegetables are rich in pharmacological substances that act in a therapeutic way and, together with fruits, are the most important source of vitamins, minerals and dietary fibre. The World Health Organization recommends a daily intake, excluding potatoes, of five servings of vegetables and fruits a day.

### Nutritional claims

Throughout the book, we have described vegetables as being an 'excellent', 'rich', 'good', 'useful', or just plain 'source' of a particular nutrient. The terms describe the nutritional value of the vegetable in relation to recommended daily allowances (RDAs), but since RDAs vary from country to country, the descriptions are a guide only.

'Excellent' refers to vegetables which provide 100 per cent of the UK RDA; 'rich' means they provide 75 per cent; 'good' 50 per cent; and 'useful' 25 per cent. A vegetable that is said to be a 'source' of a nutrient contains a minimum of 10 per cent of the RDA.

### Antioxidants

Widely distributed in plants are antioxidant nutrients, which protect the body by deactivating harmful substances known as 'free radicals'. It is believed that these free radicals attack DNA, the genetic material in the nucleus of a cell, and the resulting changes may cause cancer. They also cause oxidative damage, which has been linked to the onset of premature ageing, cataracts and hardening of the arteries.

The most common antioxidants are bioflavonoids, carotenes, vitamins C and E, and zinc, selenium, manganese and copper.

### Bioflavonoids

These occur widely in vegetables, especially the dark green leafy types and red peppers. Quercetin, one of the most well known, is found in cabbages and is believed to be a powerful anti-inflammatory, anti-viral and anti-tumour agent.

### Carotenes

Carotenes are found mainly in orange- and red-fleshed fruit and vegetables, but also in dark green leafy vegetables. As well as the well-known beta-carotene, they include alpha-carotene, cryptoxanthin, lutein, zeaxanthin and lycopene. Carotenes have diverse biological functions. Some are precursors of vitamin A and are metabolically converted to vitamin A in the body. These types include alpha- and beta-carotenes and cryptoxanthin. As well as being effective antioxidants, carotenes boost the immune system and also protect the skin from damage caused by UV radiation.

Lutein and zeaxanthin are linked to the functioning of the part of the retina which is responsible for sharp and detailed vision.

Some carotenes are associated with decreased risk of specific cancers. For example, lycopene is thought to help protect against prostate cancer; lutein, alpha- and beta-carotenes against lung cancer; beta-carotene against oral cancers; and cryptoxanthin against cervical cancer.

### Glucosinolates

Glucosinolates are found almost exclusively in cruciferous vegetables – broccoli, cabbages and kales, for example. Many research studies suggest that vegetables high in glucosinolates are associated with a reduced risk of cancer, particularly colon cancer. Other studies have identified a slight risk that excessive consumption of glucosinolate-containing vegetables could have a toxic effect, or cause goitre where there is an iodine deficiency.

### Sulphur compounds

Some studies suggest that the sulphurous compounds which give the alliums – onions, garlic and leeks – their characteristic odours may help suppress tumour formation. There is also evidence that these compounds might be particularly helpful in protecting against gastric cancers.

### Proteins

Proteins are vital for the growth, repair and maintenance of the body. They are made up of chains of amino acids, eight of which are described as 'essential' since the body cannot make them for itself. Protein from plant foods is usually deficient in one or more of these amino acids. When plant foods from different sources are combined, however, the protein quality is improved. Good examples of protein combining are serving bread or tortillas with beans, or mixing beans with rice. Most fresh vegetables are very low in proteins, but peas, beans and mashuas contain small but significant amounts.

### Carbohydrates and dietary fibre

Carbohydrates are the complex sugars and starches that provide the fuel needed to produce energy. Root vegetables, such as potatoes and salsify, are an excellent source. Dietary fibre is a form of carbohydrate known as non-starch polysaccharide, a collective term for the celluloses, hemicelluloses and pectins found in the structural material in the roots, stems, leaves, seeds and fruits of plants. It stimulates the digestive system, reduces the risk of bowel cancer, heart disease, obesity and diabetes, and helps prevent constipation.

### Vitamin A

Sometimes referred to as retinol, vitamin A is found in foods from animal sources. It is also derived from some of the plant carotenes, which convert to vitamin A in the body (see Carotenes).

### B vitamins

This is a group of almost a dozen substances involved mainly with the release of energy from food within the body. It includes eight actual vitamins, folate (the collective name for compounds derived from folic acid), and several vitamin-like compounds. B vitamins are required for the functioning of the immune system, the digestive system, the heart and muscles, and for the production of new blood cells. Some of the B vitamins have antioxidant properties.

Most vegetables contain small quantities of B vitamins, but dark green leafy types, beetroot, fennel and summer squash are particularly good sources of folate. Folate is vital for the formation of new cells and therefore for the growth of the baby in the womb and normal development in children.

### Vitamin C

Vitamin C assists with the absorption of iron from food, and with the formation of bones, teeth and tissues. It speeds the healing of wounds, helps keep the skin elastic and improves resistance to infection. It is an important antioxidant, protecting the body against the harmful effects of free radicals and thus reducing the risk of cancer. The body cannot synthesize vitamin C, so we have to get it from food. Vegetables are a major source, particularly peppers, chillies, tomatillos, leeks and dark green leafy types. Since the vitamin is easily destroyed by prolonged storage, exposure to air or heat, vegetables should be as fresh as possible, prepared just before serving and eaten raw or lightly cooked.

### Vitamin E

Vitamin E is an important antioxidant, protecting the cells from oxidative damage and the risk of cancer. It boosts the immune system, prevents muscle inflammation, and may help reduce the symptoms of arthritis. Vitamin E is also involved in red blood cell formation. Small but useful amounts are found in peas, beans, onions, peppers, cabbage and broccoli.

### Calcium

In addition to strengthening the bones, calcium is essential for muscle contraction (including that of the heart muscle), nerve function, enzyme activity and the clotting of the blood. Too little calcium in the diet results in stunted growth and rickets in young children, and osteoporosis (loss of bone) in post-menopausal women. Calcium is found in green leafy vegetables, globe artichokes, salsify and broccoli.

### Magnesium

An essential constituent of the body cells, magnesium activates energy-releasing enzymes and is required for muscle and nerve function. Studies show that low magnesium levels may be associated with increased risk of heart disease. Magnesium is found in dark green leafy vegetables, globe artichokes, potatoes, salsify and garlic.

### Iron

Iron is involved in the production of red blood cells, the functioning of several enzymes and in transporting oxygen around the body. A deficiency results in anaemia, the symptoms of which include tiredness, breathlessness, pale skin and irritability. Vitamin C promotes the absorption of iron, whereas oxalic acid (in spinach) and phytic acid (in grains) inhibit it. Iron is present in significant amounts in oriental greens, and at lower levels in leeks, garlic, salsify and Jerusalem artichokes.

### Zinc

One of the most versatile of minerals, zinc regulates enzyme activity and is involved in the metabolism of proteins, carbohydrate, energy and genetic material within the cells. It is essential for growth, especially of the baby during pregnancy, for the formation of bone tissue, the development of reproductive organs, the healing of wounds and for maintaining a healthy immune system. It is also thought to be helpful in preventing and shortening colds. Amaranth leaves are a particularly rich source. Peas, beans and fennel contain small amounts.

### Potassium

Potassium is needed for the normal functioning of the nerves and muscles, and is involved in enzyme activity and protein metabolism. It works in a complementary way to sodium (salt) in the concentration and balance of fluids within the cells. A reasonable intake will help remove excess sodium and so may be effective in preventing high blood pressure. All vegetables contain potassium, but oriental greens are a particularly rich source.

### Selenium

Although needed in very small amounts, selenium is an important trace mineral. It works in conjunction with vitamin E as an antioxidant. As such, it protects the cells from oxidative damage, inhibiting cancer and reducing the risk of heart disease. Small amounts are found in peas and beans.

## Ingredients

Some of the ingredients used in the recipes may be unfamiliar and are not to be found in every supermarket. In most cases, we have suggested an alternative, but it really is worth trying to find the more obscure ingredients so you can experience new flavours and expand your culinary repertoire. Good hunting grounds are street markets, shops catering to ethnic communities and the better health food stores. There is now also an increasing number of specialist mail order food companies supplying unusual flours, grains, spices and oils.

### Oils

It is worth investing in good-quality extra-virgin olive oil, particularly for salads. Increasingly, though, we like unrefined extra-virgin sunflower oil, now reasonably easy to find in supermarkets and health food shops. It is lighter than olive oil, has a deliciously satiny mouth-feel, and is a better choice for dressing oriental-style salads.

Groundnut oil should be used for deep-frying and stir-frying, as it has a high smoke point. Grapeseed and refined sunflower oil are good all-purpose neutral oils that are not too heavy and are good for shallow-frying.

Unrefined walnut and hazelnut oils add a wonderful depth of flavour to salad dressings, as do pumpkin seed, pine kernel and macadamia oils. These oils are rich, complex and expensive, so mix them with a neutral oil to make them go further. Toasted dark sesame oil is extremely strong and should be used more as a flavouring agent than a dressing.

### Vinegars

There is a wide choice of vinegars. White wine vinegar is lighter than red wine vinegar and goes best with mild-tasting leaves. Balsamic vinegar, made in Italy from sweet wine, is aromatic and full-bodied. A few drops go a long way. Sherry vinegar is another full-bodied type, though not as sweet as balsamic vinegar.

Cider vinegar, one of our favourites, has a low acidity and a good fruity flavour. Rice vinegar is good for oriental-style salads and mixes well with sesame oil. Chinese black vinegar, another favourite, has a mellow meaty flavour but is quite light.

### Soy sauce

We prefer the traditionally made Japanese soy sauces, such as shoyu, or tamari, a wheat-free version. They have a warmer, mellower flavour than the Chinese type, and can be bought in health food stores and some supermarkets.

### Sea vegetables

Seaweeds – or sea vegetables, as they are often called – are one of the richest sources of vitamins and minerals. They are mostly imported from Japan and are available dried in good health food shops and supermarkets. If you are new to them, try the milder varieties such as arame, hiziki or nori. They add a delicious flavour and texture to all kinds of vegetable dishes.

### Salt and pepper

Always use freshly ground black pepper. Coarsely ground peppercorns tend to look more attractive than those finely ground to a dust. Green peppercorns are sold dried or bottled in brine. They add a delicious smoky flavour to dishes, and can be used whole or lightly crushed.

When it comes to salt, we use unground sea salt flakes wherever possible. The flakes taste quite different from ordinary salt and provide crunchy little bursts of flavour on the tongue.

### Spices and dried herbs

We do not often use dried herbs, since they have little flavour, but robust types such as rosemary, thyme and oregano are good in pasta sauces and casseroles, and are sometimes even preferable to fresh herbs. Both spices and dried herbs benefit from dry-frying to bring out their flavour. Put them in a heavy-based pan without any oil and fry over moderate heat until you smell the aroma. Take care, as they burn easily.

### Stock

We wholeheartedly recommend that you make your own stock, particularly for soups. Commercial stock cubes are simply too harsh and salty. The only exception is a Swiss vegetable bouillon powder which can be used as a vegetable stock or to replace chicken stock.

# VEGETABLE FRUITS & FLOWERS

CHILLIES & SWEET PEPPERS   AUBERGINES

TOMATOES   TOMATILLOS   SUMMER SQUASH

WINTER SQUASH & PUMPKINS

EXOTIC CUCURBITS   CUCUMBERS

GLOBE ARTICHOKES & CARDOONS

BROCCOLI & CAULIFLOWER

# CHILLIES & SWEET PEPPERS

All peppers, both the sweet ones and the hot chillies, are members of the genus *Capsicum*. The word may be derived either from the Latin *capsa* (box) or the Greek *kapto* (to bite), perhaps referring to the hot pepper's habit of biting back when bitten into. The genus consists of five domesticated species: *C. annuum*, the most common, with the greatest number of varieties; *C. baccatum*, known as *aji* in South America and one of the least-known in the northern hemisphere; *C. chinense*, which includes some of the world's hottest peppers; *C. frutescens*, used in Tabasco sauce; and *C. pubescens*, with hairy leaves and black seeds.

## A vibrant palette

Those with an artistic temperament will appreciate the colours in peppers, which span the spectrum from cool greens to vibrant reds. The colours vary according to the variety and age of the fruit; and, like artists' paints, depend on the presence or absence of various pigments.

Most varieties start life as green, as a result of the presence of chlorophyll, the same pigment responsible for the green colour in leaves. Sometimes the chlorophyll is absent, however, and the fruit can be either creamy white or, owing to anthocyanin pigments, an aubergine-like purple. As the fruits mature, carotenoid pigments are produced, while chlorophyll and anthocyanins disappear, creating vivid hues of yellow, orange and red. In some varieties, however, the chlorophyll persists as the fruits ripen. If this happens alongside the production of red carotenoids, the mixture of green and red creates a brown fruit.

## Taste and aroma

The taste and aroma of peppers and chillies arise from a blend of different chemical compounds, including volatile oils and sugars. As the fruits mature, the best varieties become sweeter and develop a depth and complexity that some *aficionados* compare to that of a fine wine.

The components that receive the most publicity are the capsaicinoids, the group of compounds responsible for the pungency in chillies. The majority of these capsaicinoids are found not in the seeds and flesh, as is commonly thought, but in the placenta, that part of the fruit to which the seeds are attached. However, you have only to bite into the seeds of a chilli to realize that they, too, can be very hot. Removal of the placenta and seeds will reduce the heat considerably.

While scientists measure the capsaicinoid levels in peppers in Scoville units, using sophisticated analytical equipment, the cook's best measuring technique is simply to taste the fruit. Heat levels vary enormously and it is always wise to approach unfamiliar peppers with caution. Should you find yourself with a mouthful of fire, the most effective antidotes are dairy products or starchy foods, such as bread or rice. A drink of cold water or beer is no good at all, but will actually seem to increase the heat.

## Appearance

The shape and size of the mature fruit, while determined to a certain extent by growing conditions, are affected mainly by the genetic make-up of the variety. The names of these varieties offer clues to their shape and size. Descriptive names use animal parts, as in 'Bird's Eye', 'Rooster Spur' and 'Goat Horn', as well as the names of other fruits and vegetables, such as 'Cherrytime', 'Sweet Banana' and 'Mushroom'. Even articles of clothing are employed, as in the tam-shaped 'Scotch Bonnet', and concessions have been made to modern technology with the so-called 'Eastern Rocket'.

## Cultivation and harvesting

If you are growing your own peppers it is wise to cater to the plant's tropical pedigree and provide warm growing conditions. In cooler climates, it may be necessary to provide a sheltered spot in the garden, or even use a greenhouse or polythene tunnel.

Growing conditions are thought to exert a strong influence on heat levels, and fruits picked in the cooler autumn months may not be as hot as those picked in the summer. Heat levels also seem to increase with maturity, though they may decline if the picking is delayed until after the fruit has changed colour.

## Buying and storing

When shopping for peppers, farmers' markets and ethnic stores are rewarding hunting grounds for the more interesting varieties. Look for firm lively fruit, with vibrant colour and smooth, glossy skin. Reject any wrinkled specimens or those with brown marks or watery bruises. Fruit in prime condition can be stored for a week or two in a ventilated plastic bag in the fridge. Chilling does nothing for flavour, so bring raw fruit to room temperature before use.

## Preparation

The easiest way to prepare a whole sweet pepper is to cut a thin slice from the stalk end. If it is to be stuffed, cut away ribs and seeds. Otherwise, slice it in half lengthwise before removing them.

CHILLIES

'Rooster Spur'

'Pasilla Bajio'

'Mulato Isleño'

'Ortega'

'Early Jalapeño'

'Cherry Bomb'

'Hungarian Hot Wax'

'Super Serrano'

'Early Scotch Bonnet'

'Habanero'

To prepare chillies, slit them lengthwise, remove the seeds and ribs with the tip of the knife and cut off the stem. Rinse under cold running water and prepare according to the recipe. It is essential at this point to wash utensils and to scrub your hands thoroughly. Unless you want to experience a burning glow for hours, do not rub your mouth, eyes or face until you have cleaned up.

To roast sweet peppers and chillies, put them under a preheated very hot grill, directly in a gas flame or – best of all – over hot coals, until the skin blackens and blisters. Chillies need less time, as they tend to disintegrate if over-roasted. If the skin doesn't peel away easily, put the peppers in a sealed container and leave them for 10 minutes: the steam helps loosen the skin. Don't be tempted to rinse roast peppers under running water, or you will wash away the lovely smoky juices.

## Cooking

With their robust flavours and bright colours, sweet peppers bring life to a wide variety of dishes. Added to a slowly stewed *soffrito* of onions, they contribute to the underlying flavour of many Mediterranean dishes. Boxy bell peppers provide the perfect receptacle for stuffing with fragrant mixtures of grains, minced meat or nuts.

For a richly flavoured salad, try raw or roasted red and yellow peppers dressed with fruity olive oil, balancing their sweetness with the saltiness of olives, capers or anchovies.

Roasted and puréed peppers make vibrant sauces – try rouille, a spicy Mediterranean sauce that transforms a fish soup, or serve a multicoloured trio of purées to scoop up with good crusty bread and a platter of crisp crudités.

Although mainly known for their heat, chillies used judiciously can add pleasing flavour accents to bland dishes. A small amount of finely chopped chilli is good with grains and pulses or mildly flavoured root vegetables, or add with garlic to a simple dish of pasta dressed with olive oil and coarsely ground black pepper.

Chillies are also great for livening up white fish – chop a little into a marinade of lime juice, olive oil and coriander, and brush the fish with this while grilling.

For Chinese stir-fries, sizzle small whole chillies with garlic and ginger for a few seconds and then remove them from the pan. The flavour will permeate the oil, adding zest to the dish. Thin slivers of raw chilli or chilli tassels also make a stunning garnish.

Chillies can even be added to sweet dishes. A chilli-flavoured syrup is surprisingly good with chilled slices of tropical fruit; or try the sensational schizophrenic combination of cold hard guava ice-cream (page 22) spiked with 'Habanero' – the hottest chilli of all.

## Nutrition

Peppers and chillies are powerhouses of vitamins vital for good health. Weight for weight, an uncooked sweet pepper provides over twice the vitamin C of an orange, while red chillies potentially provide an astonishing seventy-five times as much. Red peppers are an excellent source of beta-carotene, containing twenty-five times more than yellow ones. They also contain bioflavonoids which, together with beta-carotene and vitamin C, are antioxidants widely believed to protect against heart disease and some cancers. Peppers contain a small amount of vitamin E – another antioxidant – and are a useful source of B vitamins and potassium.

## Varieties

Most of the sweet peppers and chillies we grow belong to the *annuum* species.There are hundreds of varieties to choose from, and it can be a daunting task deciding which to grow.

### Chillies

Though pungency is their most talked-about feature, there is more to chillies than just giving a kick to a dish of *chile con carne*.

The fruits of ancho or poblano types are heart-shaped, thin-fleshed and only mildly hot. Dark green when young, some varieties turn red as they mature, while others, such as '**Mulato Isleño**', turn a rich chocolate brown. This variety develops a deep, complex sweet flavour, and it is among the best-tasting of peppers. Once its tough skin is removed, the fruit can be used in various ways. It is excellent stuffed with meat or cheese in the classic Mexican dish Chiles Rellenos (page 20), or try it fried with potatoes and garlic, or marinated in olive oil and vinegar with a little chopped garlic.

The elongated New Mexican types, also referred to as Anaheim, are green when immature. They sometimes turn yellow, orange or brown as they mature, though most varieties, including '**Ortega**', turn red. As with other New Mexican chillies, 'Ortega' is only mildly hot and is used both green and red. After roasting and peeling, it is ideal for stews, salsas and *chiles rellenos*.

Jalapeños are thick-walled and torpedo-shaped. The immature fruits are normally green, with smooth skin, though purple colouring and corky striations occasionally develop. Both conditions are natural, and should in no way should detract from your eating pleasure. '**Early Jalapeño**' is a hot variety that turns red at maturity. It is usually eaten green, perhaps finely diced and mixed with onion, lime juice and fresh coriander in a salsa, or sliced into rings for pizza toppings, or served with crisp tortilla chips dripping with melted cheese.

Serrano types are thinner and more sausage-shaped than the

jalapeños. A typical variety is '**Super Serrano**', which has fruits with medium-thick flesh that are normally used green.

'**Hungarian Hot Wax**' is one of our favourite chillies for growing and eating. Though not the best-tasting chilli, it is both early-yielding and highly productive: we have counted more than thirty fruits on a single plant. It is also stunningly beautiful, with its elongated fruits changing from light green or yellow to sunset oranges and reds as they ripen. Its medium-thick flesh is not overly hot, and it can be used in much the same ways as a sweet pepper. Try it in a Mediterranean stir-fry with courgettes, tomatoes and purple onions (page 20), or sliced in a salad with juicy, sweet 'Sungold' tomatoes.

'**Pasilla Bajio**' is the most elegant chilli of all. It is dark green and furrowed, with a long graceful shape. The fruits turn a beautiful dark brown when they mature, and look like Christmas tree ornaments as they hang from the plants. They are quite mild, with thin flesh and tough skin.

The saying that big things come in small packages is definitely true for '**Rooster Spur**'. These tiny cayenne look-alikes are packed with heat and should be eaten only by the stout-hearted. Try adding them whole to stir-fries, or marinate them with sliced cabbage and lime juice in a salad.

The name '**Cherry Bomb**' is partially a misnomer. It is certainly explosive in its impact, but the shape of our fruits, especially early in the growing season, was more like a spinning top than a cherry. The fruit has rather tough skin, which should be removed. With its thick flesh, however, it is perfect for salads to accompany barbecued meats.

'**Habanero**', belonging to the *chinense* species of chilli, has ripe fruits that are salmon-orange, with a wrinkled lantern shape. The quintessential pepper from hell, it has a pleasantly fruity tropical aroma that lends a special quality to almost any dish. The fruits may not ripen until well into the summer or early autumn, but the wait is worth it. Not only are they very hot, but they are also oddly delicious when added to recipes with tropical fruit. The name 'Scotch Bonnet' usually refers to chillies belonging to the *chinense* species; '**Early Scotch Bonnet**', however, is classified as an *annuum*. It is pungent, red when mature and (no surprises here) the shape of a Scotch bonnet.

**Sweet peppers**

Although sweet peppers come in a broad range of shapes and colours, flavour differences are not so marked.

Some of the odder varieties include '**Top Boy**', a tomato-shaped pepper with very thick flesh that turns from green to yellow. Compared to other varieties, we found it good raw, though a little bit on the mild side. '**Slim Pim**' is about the size of the middle finger and has an undulating surface, thin flesh and tough skin. The raw flavour can be unpleasantly astringent, and overall it is not a particularly edible pepper. A so-called Cuban type, '**Biscayne**' has an elongated shape and medium to thick flesh. The fruits are sweet and tasty when mature, with a bright red colour.

'**Corno di Toro**' (bull's horn) perfectly describes the long curved shape of the red and yellow varieties that go under that name, and are especially popular in Italy. They are versatile peppers that can be used raw in salads, as well as fried, roasted, grilled, stuffed and stewed, lending authenticity to Mediterranean dishes.

The fruits of '**Apple**' are wedge-shaped, although they can sometimes have rounded ends. They have tender thick flesh that is juicy, sweet and mild, with a touch of tanginess. Overall, this is a satisfying pepper to eat raw, especially in salads.

Bell pepper is the generic name for fruits that are blocky in shape. The diminutive '**Jingle Bells**' is bright red, fairly thin-fleshed and tangy when ripe. It is an ideal candidate for stuffing as a bite-sized appetizer. Try the recipe for Stuffed Baby Peppers on page 22. '**Gypsy**' is a slightly elongated bell that changes from light green or yellow to bright orange and red as the fruits mature. They are early to mature and highly productive, and the flavour of their thick flesh is deliciously sweet. This is a good all-round variety that is easy to grow and has much to recommend it.

The taste of '**Islander**' leaves a lot to be desired: in immature fruit it can be unpleasantly grassy, becoming sweeter (though sometimes still slightly astringent) when red and ripe. However, the colours more than make up for the less-than-ideal flavour. When young, the fruits have a delicately ivory-coloured flesh that is a perfect foil to their shiny lavender skin. As they mature, they go through a neon-light colour show of violets, oranges and yellows, before they become a deep mature red. '**Ariane**' turns an intense orange when ripe, and has a well-balanced sweet flavour.

# Roasted Pepper Bruschetta

**Serves 4 as a snack or light meal**

*3 red or yellow peppers*
*1 small red onion, cut into 1 cm/½ inch thick slices*
*2 small courgettes, quartered lengthwise*
*olive oil for brushing*
*dash of balsamic vinegar*
*4 ciabatta rolls, halved horizontally*
*2 garlic cloves, crushed with the flat of a knife blade*
*a few rocket leaves*
*sea salt and freshly ground black pepper*
*a few leaves of basil, shredded*

**This is a blissful combination of soft fleshy peppers and craggy Italian bread. Thick-fleshed pepper varieties, such as 'Gypsy', 'Corno di Toro' or 'Apple', are best, otherwise use firm bright-looking supermarket specimens. Use cocktail sticks inserted horizontally to keep the onion rings in place while grilling.**

1. Preheat the oven to 230°C/450°F/gas 8. Put the peppers, onion slices and courgettes in a large roasting pan. Lightly brush them with olive oil and roast in the oven for 15–25 minutes, turning occasionally, until the courgettes are golden and the onion and peppers are blackened.
2. Peel away the skin from the peppers, discard the seeds and cut the flesh into thin strips. Put these in a small bowl and toss with a dash of vinegar and just enough olive oil to coat. Cut the courgette slices into bite-sized pieces and cut the onion slices in half.
3. Put the bread on a baking sheet and brush on both sides with oil. Toast in the oven until golden brown. Rub the cut sides with mashed garlic.
4. Arrange a few rocket leaves and the vegetables on the base of each roll. Sprinkle with sea salt, black pepper and some shredded basil. Place the other half of the roll on top.

## Hungarian Hot Wax Peppers with Tomatoes & Courgettes▸

**Serves 2–4 as a light main course**

*2 tablespoons olive oil*
*350 g/12 oz 'Hungarian Hot Wax' peppers, or thin-fleshed yellow peppers, deseeded and cut into 2 cm/¾ inch strips*
*3–4 small purple onions, or large spring onions, thinly sliced*
*350 g/12 oz courgettes, sliced*
*2 fat juicy garlic cloves, finely chopped*
*400 g/14 oz vine-ripened tomatoes, or 400 g/14 oz can of chopped tomatoes*
*2 teaspoons cider vinegar*
*salt and freshly ground black pepper*
*1 tablespoon chopped fresh coriander*

**This Mediterranean dish of brilliant reds, yellows and greens brings together the most delicious of the vegetable fruits. Crisp-textured, thin-fleshed 'Hungarian Hot Wax' peppers provide just enough heat to perk up the taste buds. If you don't grow your own, you can buy these peppers in Middle Eastern food shops.**

1 Heat the oil in a large frying pan and add the peppers, onions and courgettes. Stir-fry for a few minutes over moderate to high heat, until they are just beginning to soften.

2 Add the garlic. Reduce the heat a little, cover and leave for 2–3 minutes – the vegetables will cook in their own steam.

3 Stir in the tomatoes, the vinegar, salt and plenty of freshly ground black pepper. Cook, uncovered, for 3–4 minutes more. By now the vegetables should be producing plenty of juice.

4 Tip the contents of the pan into a colander set over a bowl. Allow the juices to drain, then transfer the vegetables to a warmed serving dish. Pour the juices back into the pan, bring to the boil and simmer briskly until reduced and slightly thickened.

5 Pour the sauce over the vegetables and sprinkle with the coriander.

## Chiles Rellenos

**Serves 6 as a light main course**

*6–10 New Mexican or poblano chillies, 13–15cm/5–6 inches long*
*groundnut oil for brushing and deep-frying*
*175–225 g/6–8 oz feta, Cheddar or other crumbly hard cheese*
*3 tablespoons seasoned flour*
*3 large eggs, separated*
*¼ teaspoon salt*
*Grilled Tomato Sauce (page 40), to serve*

***Chiles rellenos* (stuffed chillies) is one of Mexico's classic dishes. It's a good idea to grill more chillies than you need as there may be a few disasters when peeling and stuffing. The aim is to char the skin without cooking the flesh too much, otherwise the flesh tends to disintegrate at the stuffing stage.**

1 Lightly brush the chillies with oil and place them directly on a gas flame or under a very hot grill, turning them regularly until they are evenly blistered and blackened. Immediately transfer to a sealed container and leave them to steam for 10 minutes, then rub off the skins.

2 Cut a lengthwise slit down the side of each chilli (there is usually a weak point), stopping about 2 cm/¾ inch from the tip. Carefully scrape out the veins and seeds. Try to resist the urge to do this under running water or you will wash away the delicious smoky juices.

3 Preheat the oven to 190°C/375°F/gas 5. Cut the cheese into 8 × 1 × 1 cm/3¼ × ½ × ½ inch strips. Place these inside the chillies, then overlap the cut edges to enclose the cheese. Wipe the chillies dry with a paper towel and dust with seasoned flour.

4 Beat the egg whites until stiff, then beat in the yolks one at a time with the salt.

5 Heat enough oil in a large heavy-based frying pan to come 5 mm/¼ inch up its sides. When the oil is very hot but not smoking, using tongs, dip the chillies in the batter one at a time, then immediately transfer to the pan. Fry in batches until golden and crisp, turning once. Drain on paper towels and keep warm in the oven while you fry the rest. Serve with the tomato sauce.

## Stuffed Baby Peppers▸

**Makes 24**

*12 baby peppers, about 4 cm/1½ inches long, halved lengthwise and deseeded*
*sea salt and freshly ground black pepper*

**For the avocado stuffing:**
*1 large avocado, mashed*
*juice of 1 lime*
*1 spring onion (green parts included), finely chopped*
*2 tablespoons chopped fresh coriander*

**For the goat's cheese stuffing:**
*175 g/6 oz soft goat's cheese*
*3 tablespoons shredded basil*

**For the mozzarella, pancetta and sage stuffing:**
*115 g/4oz mozzarella cheese, very finely diced*
*75 g/2¾ oz pancetta, finely diced and fried until crisp*
*1 tablespoon chopped fresh sage*

**Bite-sized peppers, such as 'Jingle Bells', make an unusual snack to serve with pre-dinner drinks or at a barbecue. There is enough of each stuffing to fill about eight pepper halves.**

1 Preheat the oven to 230°C/450°F/gas 8. Arrange the pepper halves, cut side down, in a roasting pan, leaving plenty of space between them. Roast in the oven for 7–10 minutes, until the edges are charred. Allow to cool. Peel away the skin if it is tough.
2 Combine the stuffing ingredients separately, season each to taste and fill each pepper half with a heaped teaspoonful.

## Chilli & Guava Ricotta Ice-Cream

**Makes about 1 litre/1¾ pints**

*140 g/5 oz sugar*
*1 'Habanero' chilli, deseeded*
*4 guavas, weighing about 800 g/1¾ lb*
*strained juice of 2 limes*
*250 g/9 oz ricotta cheese*
*6 tablespoons Greek-style yoghurt*

**Making ice-cream with ricotta cheese saves all that messing around making custard, and the results are just as creamy. You can use any chilli, but the 'Habanero' is the ultimate taste experience. You may want to open the kitchen window while boiling the syrup.**

1 Put the sugar in a small saucepan with 300 ml/½ pint water and the chilli. Bring to the boil, stirring to dissolve the sugar, then simmer briskly for 5 minutes. Leave to cool, then fish out the chilli.
2 Quarter the guavas, peel and cut into chunks. Purée in a food processor or blender with the lime juice and cooled syrup. Push through a sieve to get rid of the seeds.
3 In a large bowl, beat together the ricotta and yoghurt until very smooth. Mix in the guava purée, whisking well.
4 Freeze in an ice-cream maker, following the manufacturer's instructions, or pour into a shallow container, cover with cling-film and freeze for about 2 hours, until hardening round the edges. Whisk until smooth, then freeze again.

**Variation**
Replace the guavas with the same weight of chopped mango.

# AUBERGINES

Aubergines (*Solanum melongena)* are like magicians, performing sorcery on their colours with just two pigments: anthocyanins and chlorophyll. Take their magnificent shades of purple, for example. These range from delicate violets to ominous dark shades that are almost black. The differences are explained in part by the concentrations of anthocyanins, with darker-coloured fruit containing higher levels. The concentrations are affected both by the genetic make-up of the variety and by growing conditions, with cool weather resulting in a less intense colour.

Although purple is the colour we most commonly associate with aubergines, the fruit boasts a much broader spectrum. Some varieties are a soothing green, thanks to the presence of the chlorophyll; others are snowy-white, because of low concentrations of both the purple and the green pigments. There are also daring colour combinations, such as the flamboyant purple and white striping of the more stylish Italian types.

## The bitter truth

Aubergines have a reputation for bitterness that, in many cases, is unfounded. Modern aubergine varieties grown in Europe and North America are generally not at all bitter – a factor undoubtedly contributing to their increasing popularity. West African and Thai varieties, however, are often bitter to a degree, ranging from tolerable to startling. The chemicals responsible belong to the glycoalkaloid family, also found in the aubergine's relatives – potatoes and tomatoes.

A liking for bitterness can undoubtedly be acquired, although we in the West tend to find the taste unpleasant – perhaps because of the association with childhood medicines. To the oriental palette, however, bitterness is a desirable, refreshing taste that stimulates the appetite.

## Appearance

Ranging from small squat dumplings to curvaceous elongated varieties, aubergines are as intriguing in their shapes as in their colours. There also seems to be a broad range of sizes at which a particular variety can be picked. The sizes described in seed

'Tango'
'Pingtung Long'
'Little Fingers'
'Harabegan'
'Machiaw'
'Bharta'

catalogues no doubt represent the best for maximizing yields, but larger or smaller fruits usually taste just as good.

## Cultivation and harvesting

Aubergines are heat-loving plants needing either warm summer days or the protection of a greenhouse. There are some bush varieties that do well in pots, although yields might be a little disappointing. Large, ungainly plants do best in the open ground, where they can stretch out.

The fruits are most desirable in their youth, when the seeds are still white and soft, and the skin firm and glossy. As the fruits mature, the seeds become brown and hard, and bitterness may develop. The skin undergoes distinctive transformations during the ageing process: the texture becomes hard, a dull sheen replaces the attractive gloss, and the colour changes: purple fruits turn brown, and white and green varieties become yellow. Eventually, the fruit becomes inedible, with sadly little resemblance to its younger self.

## Buying and storing

Aubergines should be purchased young, when the skin is taut and smooth. Don't buy fruits that feel light, sound hollow when tapped or have lost their glossy sheen. Aubergines are best kept at room temperature but, even then, will last only for a few days before they start to shrivel. If you need to keep them for longer, store them in a plastic bag in a cool place.

## Preparation

It is not usually necessary to peel aubergines, just trim off the stem end and prepare according to the recipe. If you're going to roast them whole, prick them first, otherwise there may be a steamy explosion in the oven.

In the olden days, cookbooks invariably recommended salting aubergines, or 'degorging' as some books quaintly describe it, to draw out the bitter juices. We rarely find salting necessary with modern varieties, although it does seem to help prevent the aubergine from sopping up copious amounts of oil when fried.

## Cooking

Since the aubergine is happiest growing in the tropics and warmer temperate areas, it features most strongly in the cuisines of these parts of the world. *Caponata* (an assertive sweet-and-sour chutney-like stew) and *imam bayildi* (aubergines baked in olive oil) are almost culinary clichés of Italian and Middle Eastern cooking respectively. Indian cooks are renowned for their imaginative use of aubergines – they appear in curries, chutneys and relishes, roasted or fried with spices, and pickled whole. Oriental cooks love them too. In Japan, for example, they are pickled in salads, or sizzled with wafer-thin slices of beef (and sometimes other vegetables) in sukiyaki.

Culinary choices for the western cook are endless. Once the mainstay of the old-style vegetarian diet, aubergines are now considered mainstream and appear on countless eclectic 'new wave' menus. Aubergines are particularly good partners for assertive ingredients such as garlic, ginger and chillies. They also team up with tomatoes, hard cheeses and pungent herbs. They are not fussy about serving temperature and do not object to being reheated.

Cubed and marinated aubergines are excellent skewered with lamb and cooked over hot coals. Alternatively, try roasting whole aubergines until charred and smoky, then purée the flesh to a silky cream with garlic, tahini, yoghurt and fresh coriander. Serve with pitta bread or crackers.

For irresistible aubergine 'oven chips', lightly brush lengthwise slices of aubergine with oil, place them in a non-stick roasting pan and roast them in a hot oven until they are golden brown and slightly chewy, then cut them into small strips. Altrnatively, leave the slices whole, smear them with pesto or hummus, roll them up and spear with a cocktail stick. Sliced grilled aubergine also makes a succulent filling for crisp filo pastry or a topping for pizza.

## Nutrition

You would expect this robust vegetable to be a powerhouse of nutrients, but sadly this is not the case. The aubergine contains protein, carbohydrate and all the usual vitamins and minerals, but in only insignificant amounts. Consequently, the fruit does not appear to have any particular therapeutic benefits. It plays a major role in the West African diet, however, and is highly regarded in Nigerian folk medicine.

## Varieties

Aubergines are often seen as the sponges of the vegetable world, adding little character of their own to a dish. While this is partly true, there are significant differences in quality between varieties, and choosing a good one will significantly improve a recipe.

'**Ova**' lives up to its name by producing pristine white fruit the shape and size of an egg. It develops brown hard seeds quite quickly, has a tough skin and a remarkably bitter taste. Another white variety, '**Tango**', has an elongated shape, a nice chewy texture and pleasing flavour redolent of mushrooms. '**Ghostbuster**' (not shown) is also white-skinned, but more oval than 'Tango'. It has a buttery, silky taste, and we recommend it highly. Try it stir-fried with red peppers and thinly sliced crunchy broccoli stems, or roast it whole and mash the flesh to a purée with garlic, lemon juice, olive oil and a dash of cayenne pepper.

Purple varieties include '**Bonica**', an oval variety with a tender skin, a juicy, chewy texture and a mushroomy flavour. '**Bharta (MHB-2)**' is a round variety with a sweet taste. It is excellent stuffed with Indian-style spices (page 30). '**Little Fingers**' has an elongated shape, with fruits that are green-fleshed and bitter when young – fortunately the flavour improves significantly as they mature. '**Bandera**' is an almost oval Italian-style aubergine with purple and white stripes. Its attractive skin is matched by its rich flavour. Grill it over hot coals and serve on slices of toasted Italian bread. A long Indian variety, '**Harabegan (MHB-9)**' has pale green skin, a firm, chewy texture and a pleasantly meaty flavour. However, the skin is tough and we found some fruits to be bitter.

Oriental types include '**Pingtung Long**' and '**Machiaw**'. Both varieties are long and thin, with white tender flesh and attractive lavender skin. 'Machiaw' is sweet and juicy, definitely one to try. 'Pingtung' is mild, with a hint of butteriness in the background. Cut them into thin strips and use in a hot-and-sour stir-fry with garlic, ginger, chilli, sugar, soy sauce and rice vinegar.

**Pea aubergines** (*Solanum torvum*) grow wild throughout the tropics. Seed is not available from seed merchants, so you will have to be content with buying the fruit from oriental food shops. Despite their bitterness (or because of it), they are popular in Southeast Asian cooking. Strip the fruits from the stalks and toss them whole into a Thai-style green chicken curry. The fruits do not soften during cooking, so they add an agreeably interesting texture to the dish.

Also popular in Southeast Asia, and available in oriental food shops in the West, is a ping-pong-ball-sized white aubergine streaked with green at the stem end that is sometimes called **apple aubergine**. It is delicious raw, with a sweetish but slightly bitter flavour. Try it in a Thai salad with prawns and a sweetish dressing flavoured with lime juice and chillies (below).

West African aubergines belong to different species and are quite unlike those seen in the West. Seeds are hard to come by, but some specialist suppliers carry a few types. The large, pale green fruits of **Gboma** (*S. macrocarpon*) become brownish red when ripe. With our western prejudices, we found them unpleasantly bitter at all stages of maturity. Strongly ribbed and striped with light green when young, '**Jaxatus Soxna**' (*S. aethiopicum*) becomes a jazzy orange red at maturity. We first taste-tested this variety at the red stage, only to experience yet more bitterness. We were later advised that the assault on the taste buds might be less vicious if tasting trials were conducted at an earlier point, so we tried some fruits just as they were starting to change colour, but we found these just as bitter.

'**N'drowa Issia**' is related to 'Jaxatus Soxna', but its fruits are smooth instead of ribbed. We tasted this variety as the fruits were changing colour and found them sweet, with a firm tomato-like texture. They would be worth trying in a robust West African style stew with mutton or goat, chillies, peanuts and spices.

## Thai Salad of Aubergines & Prawns

**Serves 4 as a starter**

*8 round white aubergines*
*juice of ½ lemon*
*4 small handfuls oriental salad leaves*
*200 g/7 oz peeled cooked tiger prawns*
*coriander leaves, to garnish*

**For the dressing:**

*1 tablespoon muscovado sugar*
*3 tablespoons warm water*
*1–2 small red chillies, deseeded and very finely chopped*
*1 tablespoon nam pla fish sauce*
*1 teaspoon soy sauce*
*3 tablespoons lime juice*

**The idea for this recipe came from Sri Owen, who kindly allowed us to include a dressing from her book *Healthy Thai Cooking*. We have used the beautiful whitish-green apple aubergines available from Thai shops. They are mildly flavoured and delicious raw, as here, but if you prefer your aubergines cooked, brush the segments with oil and grill for a few minutes. If you can't get hold of Thai aubergines, use a small ordinary purple one instead. Slice it thickly, brush with oil and grill, before cutting into bite-sized segments. For the oriental salad leaves use baby bok choy and mizuna, or other dark green leaves such as baby spinach, rocket and lamb's lettuce.**

**1** Trim and quarter the aubergines, immediately dipping them in lemon juice to prevent browning.

**2** Arrange the leaves, prawns and aubergines attractively on individual plates. Garnish with coriander leaves.

**3** To make the dressing, put the sugar and warm water in a small bowl and stir to dissolve the sugar. Add the remaining ingredients, stir again and spoon over the salad.

## Grilled Aubergine Platter with Roasted Red Peppers & Feta▸

**Serves 4–6**

*600 g/1 lb 5oz assorted aubergines, such as 'Harabegan', 'Little Fingers', 'Bharta' and long Chinese aubergines*
*olive oil for brushing*
*2 garlic cloves, roasted and peeled*
*1 red pepper, roasted and peeled (see page 16), and cut into small dice*
*juice of 1 lemon*
*1½ teaspoons ground cumin*
*½ teaspoon paprika*
*salt and freshly ground black pepper*
*55 g/2 oz feta cheese, crumbled*
*2 tablespoons chopped flat-leaved parsley or coriander*
*pitta bread, to serve*

**This dish looks lovely if you make it with aubergines of varying shapes and sizes, with differently coloured skins. However, if you don't have access to such exotica, don't be deterred: it tastes just as good made with unnamed aubergines from the supermarket. Serve at room temperature.**

**1** Prepare the aubergines as follows: cut 'Harabegan' and 'Little Fingers' diagonally into 2 cm/¾ inch slices; cut 'Bharta' – and ordinary supermarket aubergines – horizontally into 1 cm/½ inch slices; cut long Chinese aubergines into 8 cm/3¼ inch pieces, then slice each piece in half lengthwise.

**2** Brush the prepared aubergines with oil and cook on a preheated ridged grill pan until golden, turning the pieces once. They will need about 3 minutes on each side. You may have to do this in batches.

**3** In a large bowl, combine the garlic, red pepper, lemon juice, cumin, paprika, salt and pepper. Add the aubergine pieces and turn them gently to coat well.

**4** Place the coated aubergine pieces on a serving platter, arranging them attractively to show off the colours and shapes. Spoon over any pepper mixture remaining in the bowl.

**5** Scatter over the cheese and parsley. Serve with the pitta bread.

## Spiced Baby Aubergines

**Serves 2–3 as a starter, or, with rice, serves 2 as a light main meal**

*6 round or oval baby aubergines*
*1 teaspoon coriander seeds, toasted and crushed to a powder*
*½ teaspoon ground cumin*
*¼ teaspoon ground turmeric*
*¼ teaspoon cayenne pepper*
*¼ teaspoon garam masala*
*¼ teaspoon amchoor powder (optional)*
*pinch of salt*
*½ teaspoon sugar*
*3 tablespoons groundnut oil*
*2 large plum tomatoes, peeled and chopped*
*chopped fresh coriander, to garnish*

**Small dumpling or oval-shaped aubergines such as 'Bharta' are stuffed with a tasty spice mix and cooked by a combination of steaming and frying. Amchoor powder is made from dried unripe mangoes and is used as a souring agent in the cooking of northern India. If you can't get it, reduce the sugar to ¼ teaspoonful (to be added with the tomatoes), otherwise the seasoning may be a little too sweet.**

**1** Make two cross-cuts in the aubergines from the base to below the stalk to form four petals.

**2** Combine the spices, the salt and half the sugar (leave out the sugar if not using amchoor powder). Stuff this mixture into the aubergines, rubbing over the cut surfaces.

**3** Heat the oil in a high-sided frying pan which has a lid and is of a size in which the aubergines can sit comfortably in a single layer. Add the aubergines, then cover and cook over low heat, turning now and again, for 10 minutes.

**4** When the aubergines begin to look soft, add the chopped tomatoes and the remaining sugar. Cover and cook for another 5 minutes, then uncover, increase the heat and cook for 2–3 minutes to reduce the sauce. Sprinkle with chopped coriander before serving.

# TOMATOES

Perhaps more than any other crop, tomatoes (*Lycopersicon esculentum)* represent today's extremes in the food production and distribution network. On the high-tech end, there are genetically engineered varieties that have prompted much debate on the nature and role of science in the modern world. On the populist front, there has been a renewed interest in tomatoes that your grandfather used to grow, due in part to the belief that they taste better, but also because they are reminiscent of a somewhat idealized, purportedly gentler, past.

Dealing with these two extremes are a public demanding tomatoes throughout the year and the hard-pressed retailer trying to supply them. While good for the farmer, this year-round demand presents a problem for the consumer. Because ripe tomatoes are easily damaged, fruit are harvested and shipped considerable distances while still unripe. They then have to complete the ripening process off the plant, never achieving their full complement of flavours and aromas. As a result, fruits on offer in the shops are invariably and unsurprisingly bland. Ever-responsive to consumer comment, some retailers now label their more expensive tomatoes as 'grown for flavour', which has caused guffaws among gardeners and shoppers alike.

## The chemical maze

Although the idea of plant-ripened tomatoes elicits nostalgic yearnings for the simple life, the reality is very complex. Even the seemingly straightforward process of ripening is actually a complicated maze of chemical reactions not always fully understood by scientists. Fortunately the plant knows what it is doing, and ripening normally proceeds like a well-rehearsed orchestra performing a symphony, creating a perfect harmony of taste, aroma and texture.

Usually the most obvious sign of ripening is a change in colour. Immature tomatoes are green, due to the almost ubiquitous chlorophyll pigments. As maturity proceeds, chlorophyll gradually disappears and, as beta-carotene levels increase, the fruits temporarily become orange. These orange fruit are the ones most often seen in the shops. Finally, lycopenes, another type of carotene, take over and give the fruit the characteristic red colour of a fully ripe tomato.

Though the reds of the ripe fruits are attractive, the tomato seems dissatisfied with the monotony of a single colour. Learning from their more sophisticated pepper relatives, many tomato varieties take on stunning and unusual colours, putting even peppers to shame. Some varieties keep their chlorophyll as they ripen, and the fruit are unexpectedly green at maturity. Others manufacture higher than normal levels of beta-carotene that combine with lycopenes to produce spectacular orange-red fruit.

While most striking colour differences are caused by genetic mutations, the fruits can also respond, chameleon-like, to their growing environment. For example, lycopenes do not form under extremes of hot and cold, and fruits ripening in the cool days of autumn or in an overheated greenhouse end up more orange than red.

Besides changing colour, ripening fruits undergo radical changes in texture. Starting out hard, they gradually soften to a state of perfect firmness as enzymes break down the cell walls of their flesh. The enzymes' work does not stop there, though – the process continues until the fruits eventually become too soft for salads, although they will still be fine for cooked dishes.

'Shirley'
SLICING TOMATOES
'Great White'
'Brandywine'
'Yellow Stuffer'
'Pruden's Purple'
'Striped German'
'Black Krim'
'Evergreen'

## Sweetness versus acidity

Flavours also change in a ripening tomato, the most obvious being an increase in sugars and a decrease in acidity. Tomatoes are the original sweet-and-sour food, and the best-tasting fruits have a good balance of sugars and acids. High acid and low sugar levels result in tart fruits, while the reverse situation produces fruits that may be bland. The worst possible scenario is a combination of low sugars and low acidity, resulting in an insipidness that even a sprinkling of sugar and vinegar cannot disguise.

While sugars and acids are the dominant features of flavour, volatile compounds are also important contributors to both aroma and taste. Literally hundreds of these chemicals have been identified in the tomato. The roles of some remain obscure, but others have been positively connected with the characteristic aroma found in fresh tomatoes.

## Gene jockeying

New skills acquired by biochemists now allow them to inject foreign genes into chromosomes. These gene jockeys have tapped into the tomato, producing fruit that colour up normally while softening slowly. More traditional breeding methods, however, have come up with much the same thing: an extended shelf-life.

We tried some of these traditionally bred long-shelf-life varieties, sometimes sold in the shops as 'vine-ripened' or 'truss tomatoes'. They were purchased when red and firm, and stayed so after weeks of hanging around the kitchen. Although a bit on the bland side, the flavour was better than that of normal shop-bought fruits, but still no match for the best home-grown ones.

## Cultivation

From the first sight of the smooth, glossy skin to the slurp of the last mouthful, eating a perfectly ripe tomato satisfies all the senses. To improve the odds of enjoying this experience, it is well worth growing a few plants of your own.

Choosing the right variety is an absolute must, but it is also essential to provide the right growing environment if optimum acid and sugar levels are to be achieved. Restricting water and growing the plants in full sunlight increases sweetness, while adding a handful of potassium fertilizer will give acidity a boost.

To those wishing to grow their own, the tomato plant offers two choices of growth habits: determinant and indeterminant. The latter are tall growers, and need staking to keep them off the ground. They also produce numerous side shoots which need pinching out. They usually yield more than determinant types and include some of the best-tasting varieties.

Growing these indeterminant varieties, though, is a rather laborious and time-consuming activity. For those less enamoured with gardening, the choice of a determinant type is highly recommended. These have a bushy, compact growth habit, and require no pinching-out or staking. However, the fruit tends to ripen all at once, and yields may be lower.

## Buying and storing

When buying tomatoes, look for those with a vividly coloured, unblemished, taut skin and firm flesh. Sadly, varieties marked 'grown for flavour' do not always live up to expectations, so it is worthwhile shopping at a greengrocer or supermarket known for selling produce at the correct point of ripeness, or finding a good farmers' market or organic grower.

To appreciate them at their best, tomatoes should be kept at room temperature – refrigeration simply kills off flavour. Ripe fruits will keep for a day or two; under-ripe ones for up to a week.

## Preparation

Two key questions when preparing tomatoes concern peeling and deseeding. Leaving the skin intact certainly makes life easier, but stringy shreds are not pleasant to come across in a cooked dish. So, unless you are going to sieve or purée the mixture, peeling is usually necessary. To do so, place the tomatoes in a bowl, cover them with boiling water and leave them to stand for 10–20 seconds. Then drain and refresh them under cold running water. The skin should slip off easily when pierced with the tip of a knife.

On the question of deseeding, the concern is more with the juice surrounding the seeds as this may make the finished dish too watery. However, if you remove the seeds from an average-sized, thin-fleshed tomato, you will not be left with very much, so the procedure is perhaps best reserved for beefier varieties. There may also be some loss of flavour as you will lose the acidity concentrated in the jelly surrounding the seeds. In a few instances, seeds may spoil the appearance of the finished dish, but this is usually a minor consideration for the home cook, often rushing to get a meal on the table. If seeds have to be removed, slice the tomatoes in half horizontally and scoop them out.

Tomatoes don't need peeling for salads and salsas, since the skin adds colour and flavour, and also helps keep the flesh intact. The seeds will have to go if making a salsa, however, as they would otherwise spoil its appearance and texture.

## Cooking

Just as some people feel they cannot survive without lemons, others feel equally passionate about tomatoes. Think of the number of dishes in which tomatoes appear – salads, soups, sauces, pizzas, pasta dishes, risottos and curries, as well as tarts, sandwiches and sorbets. Without them, an unbelievable number of everyday staples would be wiped off the culinary map.

Simplicity is the key to success with tomato-based dishes. A tomato salad may be nothing more than thickly sliced tomatoes – preferably freshly picked – dressed with a dash of the best olive oil, a little salt and pepper and a few shredded basil leaves.

An excellent raw tomato sauce for pasta can be made with half a kilo or so of peeled and finely diced tomatoes, chopped garlic, olive oil and plenty of shredded basil, while a richly flavoured cooked tomato sauce is a matter of simmering peeled chopped tomatoes with an onion and a decadent amount of butter.

For tomato salsa, combine two or three finely diced unpeeled but deseeded tomatoes with chopped shallots or baby onions, garlic, shredded basil or chopped coriander, olive oil, black pepper and sea salt flakes. Leave it to stand for 30 minutes, but enjoy it before the freshness fades. For an Indian-style fresh chutney, mix chopped tomatoes with chillies, salt and plenty of mint.

Served with a blob of luscious garlic mayonnaise, char-grilled tomatoes are a simple but spectacular dish. Brush the cut side of halved tomatoes with oil and grill them over white-hot coals for about 3 minutes on each side, until they are slightly blackened. A sauce made from smoky char-grilled tomatoes (page 40) is excellent with root vegetables or pasta.

If you are enjoying a tomato bonanza, try making oven-dried tomatoes. The flavour becomes wonderfully concentrated, so much so that even relatively tasteless shop-bought specimens benefit from the treatment. Pack a single layer of halved tomatoes in a roasting pan, sprinkle them with olive oil, salt, pepper and muscovado sugar, then bake them in a very low oven for 1½–2 hours, until they are shrivelled but still slightly moist. Serve them as an antipasto with mint and crumbled feta cheese, or as a side dish with grilled meat or poultry. They can be kept in a covered container in the fridge for up to a week, and can also be frozen.

## Nutrition

The mainstay of the healthy Mediterranean diet, the tomato contains outstandingly high levels of lycopene, a carotene strongly believed to protect against certain types of cancer. One well-known research study at Harvard University has shown that eating fresh tomatoes and tomato products (such as tomato sauce, tomato juice and pizza) about ten times a week reduced the risk of prostate cancer by nearly 45 per cent. In animal tests, Japanese researchers found that diets supplemented with lycopene significantly suppressed the development of breast cancer. Yet another study suggests that lycopene increases the production of immune cells and may be beneficial in treating AIDS. Tomatoes are also an excellent source of beta-carotene, another carotenoid thought to have anti-cancer properties. And they contain a useful amount of vitamin C and a little vitamin E.

With all this inherent goodness, it's hardly surprising that summertime snacking on freshly picked cherry tomatoes invariably induces feelings of righteousness.

## Varieties

Generally speaking, tomatoes conveniently fit into one of three groups according to size and use: the large slicing varieties, the smaller cherry types and the drier-fleshed processing tomatoes.

### Slicing tomatoes

The larger slicing tomatoes are usually eaten raw, either alone or in salads. The ones we tried were indeterminant varieties grown in polytunnels. **'Shirley'** is an example of a modern hybrid that produces smooth red fruits about 5 cm/2 inches in diameter.

Though the hybrids have a reputation for producing bland-tasting fruits, nurture can sometimes overcome nature: when well grown, 'Shirley' can produce fruits that are definitely worth eating.

'Heritage' or 'heirloom' tomatoes are older types that are making a comeback. They have received a good press of late, although it has to be said that some of this is undeserved. They can be a challenge to grow, their size can get out of hand (we have grown specimens weighing over 675 g/1½ lb) and they can produce cracked, misshapen, grotesque fruits. The greatest insult of all is that some varieties don't even taste very good. That said, varieties well worth growing do exist.

Two that are highly recommended are '**Pruden's Purple**' and '**Brandywine**'. Both have an obvious tomato flavour with a good balance between acidity and sweetness. In contrast, there are some mild-flavoured varieties, such as '**Great White**' and '**Black Krim**', which may not fulfil everyone's dream of the ideal tomato. However, their names reflect their unusual colours, which look great in a salad and endow the fruit with a culinary value on this basis alone. '**Yellow Stuffer**' (which may not strictly be an heirloom variety) looks surprisingly like a bell pepper. Its hollow fruit has thick, pasty flesh with little moisture, and is not especially good for eating raw. When stuffed and cooked, however, it undergoes a transformation, as heat concentrates the flavours and merges them with those of the stuffing. Try a piquant mixture of couscous, fiery harissa sauce, finely diced courgette and chopped fresh coriander. Despite the low moisture content, there is enough liquid to plump up the couscous, so there's no need to soak it first.

Also well worth seeking out are the spectacular '**Striped German**', with bright reds and yellows splashed throughout the fruit, and '**Evergreen**', which is a beautiful lime-green when mature. These two varieties might be relegated to novelty status if it were not that they taste as good as they look. As it is, the combination of handsome appearance and delicious flavour makes them real winners. Together, the two varieties make the most impressive salad (page 38).

**Cherry tomatoes**

These small-fruited varieties are eaten raw, either as a snack or in salads. The term 'cherry tomatoes' is something of a misnomer, since not all of them are round – or red. We grew indeterminate varieties in polytunnels.

One of the standards for a rich tomato flavour is the red '**Gardener's Delight**'. A newer hybrid that should definitely be tried is '**Sungold**'. This bright orange native of Japan has a good tomato flavour, with exceptional sweetness, although it could sometimes do with more acidity. However, if we could grow only one tomato, this would be the one. It makes a colourful salad with bulgur wheat and plenty of flat-leaved parsley (page 106).



CHERRY TOMATOES

**'Favorita'** is a round red hybrid that is highly productive, with good texture and sweetness, though it might be improved by more acidity. Not quite in the same class of flavour as 'Sungold' and 'Favorita', '**Sun Belle**' is still a good one. Its flavour is made even more desirable by its clear yellow colour and attractive plum shape. Another novel variety is '**Red Pear**', whose name says it all. There is also a partner, '**Yellow Pear**' (not shown); both have undistinguished mild flavours. '**Yellow Currant**' and '**Red Currant**' (not shown) are small-fruited varieties, classified as *Lycopersicon pimpinellifolium*. Though pleasant to the eye, the fruits are too acid for our liking.

**Processing tomatoes**

Commonly referred to as 'plum' tomatoes, these are best for drying and processing into sauces and pastes. They are also good for frying and baking, but are not ideal for eating raw, since they are low in moisture and so less juicy than other types. We experimented with seven varieties, all determinant and grown outdoors, and found that, when eaten raw, most of them were poor in flavour and texture. However, two that are worth trying raw are '**Bellstar**' and '**Sheriff**' (not shown).

To test the varieties cooked, we sliced the fruit in half and baked them in a cool oven for 1½ hours. 'Sheriff', which has a well-balanced intense flavour, was delicious in Oven-Roasted Tomato Filo Tart (page 40). '**Italian Gold**' tasted good too, though perhaps almost too sweet. '**Roma VF**' (not shown) and 'Bellstar' are reasonable, but not in the same league as 'Sheriff'. '**Nova**' and '**Incas**' are good producers and stand up well to a colder climate, but, unfortunately, flavour does not match productivity. '**Principe Borghese**', while reasonably flavoured and a good producer, is small, seedy and thin-fleshed.

## Spectacular Tomato Salad▸

**Serves 4–6**

*4–5 assorted slicing tomatoes, such as 'Pruden's Purple', 'Brandywine', 'Striped German', 'Evergreen' and 'Great White'*
*8–10 'Sungold' tomatoes, or yellow cherry tomatoes, halved lengthwise*
*sea salt flakes*
*cracked black pepper*
*a few purple basil leaves, shredded*
*2–3 tablespoons extra-virgin olive oil*
*warm ciabatta bread, to serve*

**If you don't have the slicing tomatoes specified, use 550g/1¼ lb vine-ripened tomatoes of varying sizes. It is important that the tomatoes are at room temperature, otherwise the flavour will be less than it should be.**

**1** To show off their spectacular colouring and texture, prepare the slicing tomatoes as follows: slice 'Pruden's Purple', 'Brandywine' and 'Striped German' horizontally, and then in half if the slices are very big. Slice 'Great White' and 'Evergreen' vertically into thin segments.

**2** Arrange the slices attractively on a serving plate (white looks best, as it sets off the colour of the tomatoes) and scatter the 'Sungold' tomatoes over the top.

**3** Sprinkle with sea salt flakes, a few cracked black peppercorns and some shredded purple basil. Spoon over the olive oil and serve with warm ciabatta bread to mop up the juices.

## Tomato & Chilli Custards

**Serves 4 as a starter or light meal**

*400 g/14 oz can of chopped tomatoes or 450 g/1 lb peeled and chopped plum tomatoes*
*1 tablespoon olive oil*
*1 small onion, finely chopped*
*2 garlic cloves, finely chopped*
*1–2 fresh green chillies, such as jalapeño or serrano, deseeded and sliced*
*4 eggs, lightly beaten*
*1½ tablespoons chopped fresh coriander, plus 4 sprigs for garnish*
*salt and freshly ground black pepper*
*oil for brushing*

**This is one of those good old-fashioned recipes given a new twist by the chillies and coriander. Serve it with some interesting salad leaves and, if you like, some *salsa verde* (page 43). We find canned tomatoes nearly always of good quality, especially the Italian varieties, and preferable to poorly flavoured plum tomatoes.**

**1** Tip the tomatoes into a blender and purée until smooth. Push through a sieve to get rid of the seeds.

**2** Heat the oil in a frying pan and gently cook the onion until translucent. Add the garlic and chillies (reserving a few slivers for garnish) and fry for a few minutes more until soft. Put the mixture in the blender with the tomato purée and process until smooth.

**3** Return the mixture to the pan and simmer over moderate heat for 15 minutes, until reduced a little. Season with salt and freshly ground black pepper and allow to cool.

**4** Preheat the oven to 180°C/350°F/gas 4. Lightly oil four 150 ml/¼ pint ramekins and line their bases with oiled greaseproof paper.

**5** Mix the beaten eggs and the coriander into the cooled sauce. Pour into the ramekins. Place them in a small roasting pan and add enough just-boiled water to come half-way up their sides. Bake for 35–40 minutes, until the custards are just set.

**6** Leave to rest for a few minutes, then run the tip of a knife round the tops before turning out. Garnish each custard with the reserved slivers of chilli and a coriander sprig.

## Oven-Roasted Tomato Filo Tart▸

**Serves 4–6 as a light meal**

*1.1 kg/2½ lb plum tomatoes, halved*
*1 teaspoon muscovado sugar*
*salt and freshly ground black pepper*
*3 tablespoons olive oil, plus more for the tin*
*4 sheets of filo pastry, each 30 cm/12 inches square*
*115 g/4 oz dry goats' cheese, crumbled*
*½ teaspoon chopped fresh thyme*
*a few basil leaves, torn*

**This crisp, richly flavoured tart will have you coming back for more; in fact it's worth making two. Start roasting the tomatoes about 3 hours before you plan to serve the tart, or prepare them the day before.**

1 Preheat the oven to 150°C/300°F/gas 2. Place the tomatoes in a single layer in a roasting tin or shallow ovenproof dish. Lightly sprinkle with muscovado sugar, salt, pepper and about half the olive oil. Roast for 1½–2 hours, until they are shrivelled but still fairly moist. Increase the oven setting to 190°C/375°F/gas 5.

2 Lightly oil the base and sides of a 4 cm/1½ inch deep, 23 cm/9 inch diameter loose-bottomed flan tin.

3 Cover the filo pastry with a clean damp tea towel. Taking one sheet at a time, dab it lightly on one side with the remaining oil and lower it into the pan, pressing it up the sides. Fold in the edges before adding the next sheet. Rotate each successive sheet so that the corners are offset. Bake for 12–15 minutes, until pale golden.

4 Arrange the tomatoes in the pastry case, in overlapping circles. Season generously with salt and pepper. Top with the cheese and sprinkle with the thyme.

5 Return the tart to the oven for 15–20 minutes, until the tomatoes are heated through and the cheese is beginning to brown. Sprinkle with basil and serve at once.

## Grilled Tomato Sauce

**Makes about 600 ml/1 pint**

*1 onion*
*1 tablespoon olive oil, plus more for brushing*
*4 garlic cloves, unpeeled*
*1–2 serrano chillies (optional)*
*12 plum tomatoes, such as 'Sheriff' or 'Italian Gold'*
*½ teaspoon salt*
*freshly ground black pepper*

**The sauce may be thinned with chicken or vegetable stock if you wish. It is good with Chiles Rellenos (page 20) and grilled fish or meat. With the chillies included, the sauce perks up cooked grains or root vegetables and is excellent with tagliatelle or other flat noodles.**

1 Preheat a hot grill. Cut the onion across into thick rings about 1 cm/½ inch thick. Insert cocktail sticks from the outer edge to the centre to hold the rings in place. Brush lightly with oil and place in a grill pan with the garlic, chillies if you are using them, and tomatoes.

2 Place under the preheated hot grill, turning the tomatoes regularly and the onion rings once, until the garlic feels soft and the rest of the vegetables are lightly blackened. Peel the garlic and chillies, and remove the chilli seeds. Leave the tomatoes unpeeled.

3 Transfer everything to a blender or food processor and process until smooth. Push through a sieve, pressing with the back of a spoon to extract as much liquid as possible.

4 Heat the tablespoon of oil in a frying pan over a moderate heat. Add the purée with the salt and pepper, and cook over high heat for a few minutes, until slightly thickened.

# TOMATILLOS

The tomatillo *(Physalis philadelphica,* syn. *P. ixocarpa)* is widely grown in Mexico and Central America, where it is referred to as both *tomate verde* (green tomato) and *tomate de cáscara* (husk tomato). Tomatillos were an ingredient in Mexican cuisine even before Columbus sailed to the Americas and were probably domesticated from the wild by the Aztecs.

Sadly, the rest of the world is often deprived of this most Latin of vegetables. Availability in Britain is restricted to canned products occasionally found in speciality food shops, though these should be avoided if at all possible. Fresh tomatillos are found more easily in the US, but mostly in regions with a large Latin-American population. Cooks wishing to bring authenticity to their Mexican dishes should not despair, since seeds of a selection of varieties are readily available and the plants are incredibly easy to grow.

## A deceptive appearance

The tomatillo is something of an eccentric in the vegetable world. Both attractive to look at and excellent to eat, the fruit fits snugly inside a thin delicate husk that offers womb-like protection against the vagaries of the outside world. Removal of the husk reveals a fruit strongly resembling a green tomato, hence the frequent confusion with under-ripe tomatoes. However, appearances can be deceiving – although belonging to the same family as the tomato, the tomatillo is an entirely different species. It is, in fact, a close relative of the ornamental Chinese lantern and the delicious Cape gooseberry or physalis, both of which have similar husks surrounding their fruit.

## Cultivation and harvesting

Despite their 'south-of-the-border' associations, tomatillos are well adapted to outdoor conditions in northern latitudes. We like to make an early start with our crop, so we start our seeds indoors in spring. After the danger of frost has passed, we transplant the seedlings and grow them outdoors.

Although the plants tend to sprawl, they make up for this bad habit by sporting attractive shiny leaves and numerous purple-centred yellow flowers that gradually transform themselves into lantern-like fruit. Cultural requirements are flexible, and the plants grow well both in the garden border and, when space is limited, in large pots.

Harvesting can be quite a delicate procedure, since it requires the establishment of a degree of physical intimacy between the picker and the plant. The fruits are ready just before they completely occupy the surrounding husks, though at this stage they are impossible to see. Consequently, the fruits must be squeezed to determine their readiness for picking. As in all intimate relationships, though, this must be performed gently so that no damage is done.

Unlike many fruits, tomatillos are perfect for eating while they are still green and immature. At this point they are firm-textured and crisp, with pleasant herbaceous or grassy flavours – rather like a fine New Zealand Sauvignon Blanc. As the fruits mature, they break out of the confines of their husks and become yellower and sweeter. They are then less desirable from the cook's point of view, since the acidity decreases and the grassy flavour becomes less striking.

## Storing

Tomatillos are among those good-tempered vegetables when it comes to storage – a plus point, since they are prolific croppers. With the husks still in place, the fruit will happily keep in a covered container in the fridge for a week or two, or even three, although by this time there will be some loss of flavour. Tomatillos can also be blanched and puréed, and kept in the freezer for several months.

## Preparation

Once the husk has been removed, you'll find the surface of the tomatillo quite sticky. The uninitiated may find this disturbing, but it is a natural state of affairs soon put right by a quick rinse under running water. However, there is actually no real need to do this unless the stickiness has attracted dirt or little insects. On no account should tomatillos be peeled.

Most recipes tell you to boil tomatillos briefly or blacken them under the grill before use. We find their lovely bright flavour and firm texture are best appreciated without any cooking. However, there are some instances – in a cooked sauce, for example – when preliminary cooking is required. Grilling, or microwaving until just soft, seems to preserve the flavour best.

## Cooking

Tomatillos can be used in the same sorts of dishes as red tomatoes, although they produce less juice. They make wonderful bright-green uncooked salsas, relishes and pickles, as well as clean-tasting cooked sauces.

They are vital to the classic *salsa cruda* (raw sauce), an essential component in a Mexican meal. Put about ten tomatillos in a food processor with a chopped small white onion, one or two garlic cloves, chillies to taste, lots of fresh coriander, a pinch of sugar and salt to taste. Process to a chunky purée with a bit of water or lime juice, and use as a zappy condiment.

Another well-known classic, *salsa verde* (green sauce), is made in the same way except that the tomatillos are first cooked either by grilling or by simmering for a few minutes. For a slightly different flavour, try Tomatillo, Coriander and Coconut Chutney, or use the tomatillos like tomatoes to make Chilled Tomatillo Gazpacho (both page 44).

## Nutrition

Tomatillos are so green and clean that they have to be good for you. They are indeed rich in carotenes, are a good source of vitamin C, and also contain a useful amount of iron.

## Varieties

Tomatillos vary in a number of characteristics, including fruit maturity, size and colour.

'**Toma Verde**' produces good-sized fruit about 3 cm/1¼ inches in diameter at the green stage, when it is ready to use. It is a reliable variety, giving good yields even when grown under unfavourable conditions.

A second variety, '**Purple**', produces (unsurprisingly) fruit with purple skin. Unfortunately, the development of the necessary pigment requires light, and the husks need to be removed to produce the best colour. If any of the husk is left around the fruit, the skin develops purple and green blotches, while any purple pigment produced can leach into the cooking water. Besides purple fruit, this variety has purple-tinged stems and husks, which make it more desirable as an ornamental than 'Toma Verde'. Some of our friends are emphatic about its superior flavour.

Another variety is '**De Milpa**' (not shown), which, according to the pitch given it in the seed catalogue, is grown in the Mexican family's corn patch – *milpa* meaning 'corn field' in the local Spanish. We do not particularly like this variety, which matures rather later than either 'Toma Verde' or 'Purple', and produces disappointingly small fruit.

# Chilled Tomatillo Gazpacho ▸

**Serves 4**

*900 g/2 lb fresh tomatillos, husks removed*
*3 green peppers, halved and deseeded*
*10 spring onions, chopped*
*2 mild fresh green chillies, such as Anaheim*
*1 large garlic clove, unpeeled*
*2 tablespoons lime juice*
*40 g/1½ oz fresh coriander, stalks removed and leaves coarsely chopped*
*500 ml/18 fl oz home-made chicken stock*
*salt and freshly ground black pepper*
*1 teaspoon sugar, or to taste*
*4 tablespoons thick yoghurt*
*tortilla chips or corn tortillas, to serve*

**This is a refreshing soup with a slightly sharp grassy flavour. Use tomatillos that are firm and green, so that you can enjoy their clean bright taste. If the tomatillos are free of dirt and insects, there is no need to wash off the sticky coating.**

**1** Preheat the oven to 230°C/450°F/gas 8. Set aside four of the tomatillos, half a green pepper, four spring onions and one chilli. Coarsely chop the remaining tomatillos, green peppers and spring onions.

**2** Roast the garlic and the remaining chilli in the hot oven for 10–15 minutes, until the chilli begins to blacken and blister and the garlic feels soft. Remove the skin and seeds from the chilli and the skin from the garlic.

**3** Put all the chopped vegetables in a food processor or blender with the roasted chilli and garlic, 1 tablespoon of the lime juice and all but 2 tablespoons of the chopped coriander. Purée for 2–3 minutes until smooth, scraping down the sides often. Push the mixture through a sieve, pressing with the back of a spoon to extract as much liquid as possible. Discard the debris.

**4** Add the stock to the strained liquid. Season with salt and pepper, and sugar to taste. Cover and chill thoroughly.

**5** Just before serving, finely dice the reserved tomatillos, green pepper, spring onions and chilli. Toss in a small serving bowl with the remaining coriander and lime juice. Add salt and a pinch of sugar to taste. Taste the soup and add more lime juice or seasonings as you think fit.

**6** Ladle the soup into bowls and swirl in a blob of yoghurt. Serve with the chopped vegetables and tortillas or tortilla chips.

# Tomatillo, Coriander & Coconut Chutney

**Makes about 300 ml/½ pint**

*10–12 fresh tomatillos*
*50 g/1¾ oz fresh coriander, stalks removed and leaves coarsely chopped*
*75 g/2 ¾ oz creamed coconut, crumbled*
*1 garlic clove, crushed*
*1–3 fresh green chillies, such as serrano, deseeded and chopped*
*finely grated zest and juice of 1 lime*
*½ teaspoon sea salt*

**Tomatillos are often teamed with avocados, but try coconut for a change. The flavour is subtle, providing the sweetness that tomatillos sometimes need, but without overwhelming their fresh grassy flavour. Use as a dip with tortilla chips or crackers, or as a relish with grilled fish or poultry. Serve at room temperature as the chutney solidifies when chilled.**

**1** Place the tomatillos under a preheated grill until the skins start to blacken. Leave to cool.

**2** Tip them into a blender with their cooking juices and skins. Add the remaining ingredients and process, slowly at first, to a chunky purée, scraping down the sides of the blender quite often. If you like, add extra salt or lime juice to taste.

# SUMMER SQUASH

Unlike winter varieties, summer squash *(Cucurbita pepo)* are eaten whole – including the skin and seeds – when they are young and tender. American and Mediterranean cooks have long delighted in these delicious summer fruits, but Britain is a relative newcomer to the squash scene. The best known are the elongated green or yellow ones called courgettes or zucchini.

## Flower power

Summer squash bear deep yellow, trumpet-like flowers, pounced on by cooks for stuffing and/or deep-frying in batter (page 48). Male and female flowers grow on the same plant, with the males (identifiable by their non-fruiting green stalk) appearing first. Female flowers make a late entry and are easily recognized by the undeveloped miniature fruit to which they are attached. They are best cooked with the fruit still firmly in place.

## Buying and storing

Choose firm small squash that feel heavy for their size. The tender skin may have a few nicks here and there, but it should not be wrinkled or have any brown patches. Store squash, loosely covered, in the fridge or a cool larder for 2-4 days, depending on age. Squash blossoms should be used the day they are picked.

## Preparation

Rinse summer squash just before using them. If the skin is slightly hairy, a quick soak in cold water will float away any clinging dirt. Trim the ends, but leave the skin intact as it adds taste and texture.

## Cooking

Cook summer squash quickly and simply to preserve their delicate flavour. Steaming, grilling or lightly sautéing are the order of the day. Take care not overcook squash or they will quickly become soft. Should this happen, turn them into a chunky dip with olive oil, lemon juice and chives.

A colourful mixture of baby yellow patty pans and finger-sized dark green courgettes is excellent with tomatoes, sweetcorn and green beans in a lightly cooked summer stew. Alternatively, tip the mixture into a shallow dish, top it with grated cheese and bread crumbs mixed with thyme or finely chopped rosemary, and brown it under the grill. Another favourite is coarsely grated green and yellow courgettes briefly cooked with a generous knob of butter, some finely chopped fresh rosemary and plenty of seasoning. Serve this as a side dish with fish or poultry, or toss it with pasta shapes and freshly grated Parmesan cheese, or even turn it into a risotto.

For a wonderful warm sloppy vegetable salad, blanch wide ribbons of courgettes and carrots (pared off with a swivel peeler), and toss with some walnut oil, walnuts and chives. Alternatively, dip slices cut lengthwise into a light egg batter and deep-fry these in hot oil. Stack them on a serving plate and serve with wedges of lemon. Simpler still, steam halved small courgettes, pat them dry and smear them with pesto sauce. Both are delicious with drinks or served as a starter.

## Nutrition

Since they are over 90 per cent water, summer squash contain relatively low levels of most nutrients and they are also low in calories. However, they do provide reasonable amounts of carotene, folate (one of the B vitamins) and vitamin C, and the skin and seeds provide some fibre.

## Varieties

One of the most delicious courgettes is '**Gold Rush**' – a bright yellow variety with a crisp texture and a flavour deliciously redolent of mushrooms. It is particularly good with 'Hungarian Hot Wax' peppers and tomatoes in a colourful stir-fry (page 20). '**Costata Romanesca**' is an elongated Italian green variety. The fruit are strongly ribbed and have a superb earthy, mushroom-like flavour. This variety is remarkable for its good looks, particularly when sliced across into rounds at an angle, and the way it holds its own when cooked. It is delicious sautéed with garlic and rosemary and served on crisp polenta crostini (page 48).

Typical of the *cousa* or Middle Eastern squashes is '**Clarion**'. This is a pale green variety, short and bulbous on the blossom end and ideal for stuffing. Try filling it with a garlicky mix of breadcrumbs and mushrooms, or chopped pine nuts and bulgur wheat moistened with tomato.

'**Yellow Crookneck**' is a distinctly bulbous squash with a curved neck. Its pale yellow skin becomes warty and darker-coloured as the fruits mature. Smaller fruits have a crisp texture and a clean, faintly lemony flavour. With a good home-made stock, it makes an excellent soup.

Companions to the crooknecks are the straightneck squashes. '**Seneca Prolific**' (not shown) is elongated and slightly bulbous, with smooth yellow skin and good texture and flavour.

'**Peter Pan**' is a light green flattish scallop-edged patty pan type with a nice crisp texture but a rather mild flavour. The bright yellow '**Sunburst**' is not nearly as flat as 'Peter Pan', and is ideal for stuffing. It is well worth growing, producing an abundance of fruit of a good texture and rich flavour.

Round varieties include '**Ronde de Nice**' (syn. '**Tondo di Nizza**'). Fruits are a mottled light green in colour, with a subtle but quite meaty flavour. The dark green '**Leprechaun**' has crisp, crunchy skin and the flavour is sweet, though a little mild. It is good sliced, lightly steamed and tossed with garlic chives or shredded basil leaves and a little finely grated lemon zest. Small ones can be stir-fried either halved or whole, while the larger ones are perfect for stuffing.

## Sautéed Summer Squash with Polenta Crostini▸

**Serves 4 as a starter or light meal, or as a side dish without the polenta**

*500 g/1 lb 2oz small green and yellow summer squash*
*1 tablespoon olive oil*
*1–2 large garlic cloves, thinly sliced lengthwise*
*12–15 tender rosemary leaves*
*sea salt and freshly ground black pepper*
*finely grated lemon zest*
*rosemary sprigs, to garnish*

**For the polenta crostini:**

*115 g/4 oz polenta*
*½ teaspoon salt*
*olive oil, for brushing*

**Here you need squash with firm, meaty flesh such as 'Costata Romanesca' or 'Peter Pan'. Don't crowd the pan — fry the squash in batches, if necessary. The polenta can be prepared ahead and grilled at the last minute.**

**1** Put the polenta and salt in a saucepan and stir in 450ml/¾ pint water. Bring to the boil, stirring all the time. Simmer for about 10 minutes, still stirring, until the mixture comes away from the sides of the pan. Pour into a roasting tin about 18 cm/7 inches square, levelling the surface with a wet palette knife, and leave to cool. When ready to serve, cut the polenta into neat rectangles or triangles. Brush with oil and place under a preheated hot grill for about 12–15 minutes until golden and crisp, turning half-way though. Set aside and keep warm.

**2** Cut elongated squash into 1 cm/½ inch diagonal slices. Slice patty pans vertically, in half if they are very small, or in thickish slices if slightly bigger.

**3** Heat the oil in a heavy-based non-stick frying pan (or a ridged cast-iron one). Add the squash in a single layer and fry over moderate to high heat for 3 minutes, until the underside is lightly flecked with brown. Turn them carefully with tongs, add the garlic and rosemary and fry for 2–3 minutes more, taking care that the garlic does not burn. Remove from the heat and sprinkle with sea salt, a grinding of black pepper and a little bit of lemon zest.

**4** Place two pieces of the polenta on each individual plate, add the squash and garlic, and garnish with a rosemary sprig.

## Ricotta-Stuffed Courgette Flowers

**Serves 4 as a starter or light meal**

*8–10 courgette flowers*
*75 g/2fl oz plain flour*
*groundnut oil for deep-frying*
*salt*

*Grilled Tomato Sauce (page 40), to serve*

**For the stuffing:**

*225 g/8 oz ricotta cheese*
*1 egg, beaten*
*1 tablespoon flour*
*3 tablespoons freshly grated Parmesan*
*1 tablespoon snipped garlic chives*
*1 tablespoon chopped fresh basil*
*¼ teaspoon freshly ground black pepper*
*salt*

**If you are sure your flowers are from an organic source, there is no need to wash them — just shake them free of insects. Otherwise dip them briefly in cold water and drain on paper towels.**

**1** Carefully insert your finger and thumb into each flower and nip off the stamen. Snip off any stalk that is still attached, as well as the sepals if they are prickly.

**2** Make the stuffing by combining all the ingredients. Insert a teaspoonful into each flower and twist the petals round to keep the stuffing in place.

**3** Make a batter by pouring 175 ml/6 fl oz water into a shallow dish and gradually whisking in the flour through a sieve, making sure there are no lumps. The batter should have a consistency similar to thick pouring cream – add more flour or water if necessary.

**4** Pour enough oil into a medium-sized frying pan to come to a depth of 2 cm/¾ inch. Heat until almost smoking, then quickly dip a flower into the batter, swirling until thoroughly coated, and then into the pan. Fry for 2–3 minutes on each side, until the batter is crisp and golden. Remove with tongs, shaking to remove excess oil, and drain on a paper towel. Repeat with the remaining flowers. Sprinkle with salt and serve at once, with the tomato sauce.

# WINTER SQUASH & PUMPKINS

Botanists see winter squash and pumpkins *(Cucurbita pepo, C. moschata, C. maxima, C. mixta)* as four different species; non-scientists treat them as two different vegetables. They are, however, really one and the same vegetable, distinguished by hollow, hard-skinned fruits with yellow or orange flesh. For the sake of convenience, we call the round, flat-bottomed, smooth, orange-skinned varieties 'pumpkins'; occasionally, a few white- and grey-skinned varieties may also be referred to as pumpkins if they otherwise fit the pumpkin profile. Anything else is described as winter squash.

Whenever we grow these beauties, we marvel at their diversity. The fruits range from palm-sized chubby Tom Thumbs to monsters so heavy that they leave their impression in the soil. Their colours are hypnotically beautiful – fluorescent oranges, leafy greens and holiday tans. Some are shaped like spinning tops, some like acorns, while others have long, swan-like necks. Flesh colours vary from creamy yellow to deep glowing orange; the texture can be smooth or stringy, the flesh moist or dry, and the flavour sweet or savoury.

## Cultivation and harvesting

Winter squash and pumpkins should be planted outside only when the danger of frost has passed. They need most of the summer to grow, and are ready for harvesting in late summer or early autumn. They should be left in the sun for a week or so after harvesting, to allow the skin to toughen and dry.

The plants are notorious for the triffid-like growth of their vines. There are, however, less unruly semi-bush or short-vined varieties which would fit comfortably in a small garden, so those with limited space available need not deprive themselves of these useful vegetables.

## Buying and storing

When buying winter squash or pumpkins, look for hard, heavy fruits with no blemishes; a warty skin, however, is fine. Provided the skin is undamaged, whole fruits can be stored for several months in a frost-free, cool place.

## Preparation

Removing the tough skin is laborious but necessary for fruits that are destined for a stir-fry, soup or casserole. If you are baking or roasting large segments, however, the skin can be left in place until after cooking, when it is more easily removed .

Pumpkin recipes frequently specify a quantity of ready-made purée.The least wasteful method is to place unpeeled chunks of pumpkin in a covered dish and bake them in the oven. The flesh is easily scraped away from the skin and seeds, and this can then be mashed to a rough purée. Push it through a sieve to get rid of any fibrous residue.

Alternatively, pack peeled cubed pumpkin flesh into a large saucepan with just a splash of water. Cover and cook over a low heat until the pumpkin is tender, shaking the pan at first to prevent sticking. Tip the cooked flesh into a sieve set over a bowl and leave it to drain for a few minutes, then purée and sieve it.

'Sweet Mama'

## Cooking

Winter squash are wonderfully user-friendly. They can be baked, roasted, stuffed, puréed or fried, used in sweet or savoury dishes, and served as either an accompaniment or a main dish.

'Delicata'
'Uchiki Kuri'
'Carnival'
'Baby Bear'
'Gold Nugget'
'Rolet'
'Jack Be Little'

One of our favourite ways with squash is to roast chunky segments with olive oil, butter, a sprig of rosemary and plenty of black pepper – delicious with roast pork or turkey. Also immensely pleasing are small round squash or baby pumpkins, stuffed and baked until softly steaming and fragrant (page 54).

The puréed flesh of squash or pumpkin makes an excellent filling for ravioli, or try it instead of mashed potato in gnocchi; or work it into a yeast dough for the most deliciously moist and colourful bread.

Dense-fleshed varieties can be used in casseroles, stir-fries and sautés. West Indian cooks make a hefty stew with squash (though they call it pumpkin), beans, chillies and sweetcorn; while the Japanese dip wedges into tempura batter and deep-fry them. Also good are thin crescents fried with sliced mushrooms (page 54); or cubes fried in a little olive oil until slightly browned. Add thinly sliced garlic, a pinch of dried chilli flakes and toss for a few seconds before showering with fresh coriander.

Pumpkins can be used in much the same ways as other types of squash. Remember, though, that the flesh tends to be watery and disintegrates easily, so cooking methods should be chosen to compensate for this.

If squash vines are taking over your garden, why not follow the example of Nepalese cooks and use the shoots in a curry? Cut a 30 cm/12 inch length from the tip of the vine, discard the curly tendrils and tough stalks, and roughly chop the buds and small leaves. Heat 2–3 tablespoons of oil with some mustard seeds, fenugreek, a finely diced red chilli and a few chopped peanuts. Throw in the shoots and cook for 5 minutes with the lid on, then add a peeled, chopped tomato and cook for another 2–3 minutes – delicious with mango chutney and a mound of fluffy white rice.

## Nutrition

Like most orange-fleshed vegetables, winter squash and pumpkins contain high levels of carotenes. An average-sized serving of cooked pumpkin provides just over the recommended daily requirement, while a chunk of butternut squash provides four times that. Both contain a small but useful amount of vitamin C, as well as potassium, magnesium, iron, vitamin B6, vitamin E and folate.

## Varieties

There is an abundance of winter squash and pumpkins listed in the catalogues, and making a choice among them can be hard. It's simplified for us by our preference for the smaller-fruited varieties that can be wolfed down at a single sitting. Except where specified otherwise, the varieties mentioned below are all trailing types.

We conducted tastings in the late autumn, after storing the fruits in a shed for a few weeks. We baked them whole and ate the flesh unadorned, straight from the shell.

'**Baby Bear**' and '**Triple Treat**' (not shown) are smaller pumpkins with typically stringy orange flesh that does not taste particularly good without some kind of embellishment. Both, however, are excellent when baked with eggs, cream and spices in a traditional pumpkin pie. They produce 'hull-less' seeds with a paper-thin coat – a bonus for those who like pumpkin seeds as a snack. They are also good sprinkled over salads and soups.

Two attractive miniature pumpkins are the orange-skinned '**Jack Be Little**' and the white-skinned '**Baby Boo**' (not shown). The flesh of 'Jack Be Little' is yellow-orange, very sweet, with an agreeably sticky texture. 'Baby Boo' is light yellow with a sticky-to-dry texture and a sweet, honey-like flavour.

'**Delicata**', a short-vined variety, and '**Sweet Dumpling**' (not shown) have beautiful skin striped with ivory and green. 'Delicata' is elongated, with smooth-textured moist flesh. The flavour is almost excessively sweet, tasting distinctly of honey. The round, squat 'Sweet Dumpling' has similarly textured flesh, though it is less sweet than 'Delicata'.

We found the green, tennis-ball-sized '**Rolet**', sometimes sold under the name of 'Little Gem', quite undistinguished and bland, and '**Gold Nugget**', a bush type, is no better. '**Carnival**' falls in the same category, though it is just about saved by a touch of sweetness. It is also redeemed by its flamboyantly handsome skin – with the flesh hollowed out, this makes a terrific bowl for pumpkin soup.

The short-vined '**Sweet Mama**' has dense orange flesh that is smooth and slightly dry. Its lovely chestnut flavour is enhanced by exactly the right amount of sweetness. '**Uchiki Kuri**' has smooth-textured deep orange flesh that is just sweet enough to add some interest. Both are good for sweet or savoury dishes.

# Squash, Sweetcorn and Bean Soup

**Serves 6**

*3 tablespoons olive oil*
*1 tablespoon paprika*
*2 red onions, finely chopped*
*2 garlic cloves, finely chopped*
*450 g/1 lb peeled and deseeded squash or pumpkin (about 1.25 kg/2¾ lb unpeeled), cut into 2 cm/¾ inch cubes*
*400 g/14 oz can of chopped tomatoes*
*700 ml/1¼ pints chicken or vegetable stock*
*400 g/14 oz lightly cooked fresh shelling beans or 200 g/7 oz dried black beans, soaked, cooked until soft and drained*
*salt and freshly ground black pepper*
*225 g/8 oz frozen sweetcorn kernels or the kernels from 2 large ears of corn*
*6 tablespoons chopped fresh coriander*
*soured cream, to serve (optional)*

**Squash, sweetcorn and beans form a kind of holy trinity in the cooking of Central America. The combination is uniquely satisfying and packed with nutrients. This recipe, adapted from one in *Jane Grigson's Vegetable Book*, is delicious served with corn tortillas or freshly made corn bread.**

**1** Gently heat the oil with the paprika in a large saucepan. Add the onions, cover and cook over moderate to low heat for 10 minutes.

**2** Add the garlic, the squash, the tomatoes with their liquid and 150 ml/¼ pint of the stock. Bring to the boil, then simmer gently with the lid slightly askew for 10–20 minutes, until the squash is just tender. The cooking time will depend on the variety – if it is watery, it will soften up quite soon.

**3** Purée 3–4 ladlefuls of the soup in a food processor, then return this to the soup remaining in the pan and cook for 5 minutes more.

**4** When the squash is reasonably soft, add the beans and pour in the remaining stock. Season generously with salt and pepper and simmer for 15 minutes. Add the corn and coriander, and cook for another 5 minutes, until the corn is tender. Ladle into bowls and swirl in a spoonful of soured cream if you wish.

## Baked Squash with Couscous & Walnut Stuffing ▸

**Serves 6 as a light meal**

*6 small round squash or baby pumpkins*
*140 g/5 oz couscous*
*1 tablespoon tomato purée dissolved in 300 ml/½ pint just-boiled water*
*100 g/3½ oz small fresh or frozen broad beans*
*5 tablespoons olive oil*
*1 onion, finely chopped*
*2 garlic cloves, finely chopped*
*2 small fresh red chillies, deseeded and finely chopped (optional)*
*50 g/1¾ oz walnuts, roughly chopped*
*2 small courgettes, finely diced*
*4 tablespoons chopped fresh coriander*
*½ teaspoon salt*
*freshly ground black pepper*
*85 g/3 oz feta or dry goat's cheese, diced*
*Grilled Tomato Sauce (page 40), to serve*

**The chillies add only a touch of heat to the stuffing, which is, in any case, mellowed by the couscous. You could use a finely diced half red pepper if you prefer.**

**1** Preheat the oven to 200°C/400°F/gas 6. Trim the base from the squash so that they stand upright. Cut a slice off the top to expose the seed cavity. Scoop out the fibres and seeds.

**2** Put the couscous in a shallow dish and cover with the tomato purée solution. Leave to soak for 10 minutes.

**3** Plunge the broad beans into a pan of boiling water. As soon as the water comes back to the boil, drain immediately and slip off the outer skins.

**4** Heat 3 tablespoons of the oil in a frying pan. Add the onion and fry over medium heat until translucent. Add the garlic, chillies, walnuts and courgettes, and fry for 2 minutes more.

**5** Stir the couscous into the onion mixture, along with the broad beans, coriander, salt and pepper. Remove from the heat.

**6** Spoon the stuffing into the squash, packing it in well and piling it up. Place the cheese on top and drizzle with a bit more oil.

**7** Place the squash in an oiled baking dish and cover with foil. Roast for 30 minutes. Remove the foil and increase the oven setting to 230°C/450°F/gas 8. Cook for 10–15 minutes more, until the cheese is browned. Serve the squash with the tomato sauce.

## Butternut Squash, Shiitake Mushrooms & Spinach

**Serves 4 as a light meal**

*1 butternut squash, weighing about 550 g/1¼ lb*
*5 tablespoons olive oil*
*½ teaspoon black peppercorns, coarsely crushed*
*½ teaspoon coriander seeds, coarsely crushed*
*sea salt*
*200 g/7 oz shiitake mushrooms, thinly sliced*
*2–3 tablespoons stock*
*2 good handfuls of baby spinach, stalks removed*
*garlic croûtons (page 126)*
*chopped fresh coriander, to garnish*

**This is a simple but filling autumnal dish of contrasting textures and temperatures — warm buttery squash and mushrooms on cool greens, with a crunchy topping.**

**1** Peel and deseed the squash. Quarter the rounded part and slice lengthwise into thin crescents. Halve the neck section lengthwise, then slice into thin semicircles.

**2** Heat the oil with the peppercorns and coriander seeds in a large frying pan over moderate to high heat. Fry the squash slices in a single layer, in batches, turning with tongs, until lightly browned. Sprinkle with sea salt. Using a slotted spoon, remove to a large sieve set over a bowl.

**3** Fry the mushrooms for 5 minutes, adding some of the oil drained from the squash. Sprinkle with salt and add them to the squash.

**4** Return any drained oil to the pan. Add the stock and bubble down for a few seconds.

**5** Arrange the spinach on individual plates. Pile the squash and mushrooms on top, then pour over the pan juices. Scatter over a few croûtons (make sure they are warm), then the chopped coriander. Serve at once.

# EXOTIC CUCURBITS

Introductions to non-indigenous vegetables are more often than not pioneered by immigrants. Vegetable seeds from the 'old country' make welcome travelling companions, and the simple act of sowing them not only reinforces the link with the country left behind but is also a way of literally putting down roots in a new country. Vegetables that cannot be grown are imported and are usually available in shops located in the various ethnic communities. Eventually, these vegetables break out of their ethnic confines, find more secure places in the seed catalogues and the mainstream supermarkets, and become an intractable part of the everyday eating experience of the host culture.

Among the vegetables just starting to make a name for themselves in the West are the exotic edible gourds. They belong to the cucurbit family, and are found in shops catering to immigrants from Southeast Asia and the Indian subcontinent. They are splendidly eccentric in appearance, with fruits ranging from bitter melons – so wart-encrusted they resemble some kind of prehistoric monster – to mad twisting bottle gourds the length of an oversized wind instrument. Some are smooth and strokeable, while others are bristling with designer stubble. Skin colours include luminous limes and deep forest greens, sometimes marked with spots, sometimes striped, and sometimes neither.

The process of getting to know these exotic cucurbits is eased by the relationship western cooks and gardeners already have with the cucurbit family – courgettes and cucumbers make a regular appearance on the menu, while pumpkins show up every Halloween. Despite their strange names and appearances, the exotic branch of the family is not so very different.

## Terminology

As is often the case with vegetables, and particularly with those introduced from afar, confusion arises over terminology. Although we refer to these vegetables as exotic cucurbits, they are also called 'melons' and 'gourds' – the latter term not to be confused with that used to describe an inedible hard squash. To add to the confusion, the individual vegetables themselves are known by different names. The Chinese bitter melon, for example, has a whole string of them, including karela, balsam pear and bitter gourd; the angled loofah is variously known as silk squash, Chinese okra and gisoda.

## Cultivation and harvesting

We have had variable results growing exotic cucurbits, and we still have much to learn. In deference to their tropical origins, all of them except the cucuzzi, an Italian bottle gourd, were grown in a polytunnel. As would be expected from members of this family, the plants have a viny growth habit and need to be trellised off the tunnel floor. Because of their short vines, tindas and fuzzy gourds were easiest to manage. Perhaps the most unruly were the bottle gourds, whose undisciplined vines took over the tunnel. The cause of our problems was our lax management style: the side-shoots and main growing points were not pruned out in any of the plants, and in retrospect we can see that this was perhaps unwise.

As is typical of the cucurbit family, the gourds have separate male and female flowers, and require insects or a similar agent for pollination. There were plenty of insects in the tunnel during flowering, and reasonable numbers of fruits were subsequently harvested from most of the plants. The bottle gourds were an exception, however, producing virtually nothing; it was not until we started to hand-pollinate them that fruit production picked up. It is, in fact, a good idea to hand-pollinate all these vines. The pollen should be transferred with a paintbrush from the male to the open female flower.

The cucuzzis did well outdoors, stretching out their vines all over the ground. Fruiting was later than summer squash, but a generous yield was more than adequate compensation for this tardiness.

## Buying and storing

A year-round supply can be found in street markets and shops catering for the Indian and Chinese communities, and the mainstream supermarkets now regularly stock bitter melons, bottle gourds and angled loofahs. For some reason, smooth loofahs remain elusive.

Without exception, exotic cucurbits are best bought young, while the skin and flesh are tender. Look for small or medium-sized loofahs (20–30 cm/8–12 inches long). A fresh one will feel firm and should remain intact if you hold one end and wave it about. Tindas, snake gourds and bottle gourds should feel heavy and have unblemished skin. If they are young, the skin should be tender enough to pierce with your fingernail and the flesh should give slightly when pressed. Fuzzy gourds are best bought when 15–20 cm/6–8 inches long and still relatively downy. They should feel firm and heavy, and the skin should not have any brown markings.

If you are a newcomer to bitter melons, look for pale green stubby types with rounded warts. They are readily available in Thai or Chinese shops. There are also bright green, pointed varieties usually sold in Indian shops and now stocked by the larger supermarkets. Whichever type you end up with, they should feel firm and the skin should not have any soft patches or bruises.

Depending on condition when bought, most gourds survive well if stored, unwashed, in the salad drawer of the fridge. They are therefore worth buying even if you cannot experiment with them right away. In our experience, the most perishable are angled loofahs and bitter melons, which are best used within 2–3 days of purchase. Tindas, bottle gourds and fuzzy gourds are more robust and can be kept for a week or more. Snake gourds will keep for even longer, either in the fridge or in a cool cellar or larder.

## Preparation

Exotic cucurbits are prepared in much the same way as courgettes, marrows and cucumbers: peel them if the skin is tough, otherwise don't bother. In most cases there is no need to remove the seeds as long as the fruit is young and fresh. Make sure you wash the fruits well, or even scrub them if they are hairy and are intended to be eaten unpeeled.

Depending on the recipe, tindas, snake gourds, bottle gourds and fuzzy gourds can be thickly sliced into rounds, cut into wedges or cubes, or hollowed out and stuffed. Despite appearances, the ridged skin of the angled loofah is usually tender enough to be eaten. Just shave off the tips of the ridges, then slice the loofah at an angle into thick ovals to show off its intriguing profile.

Bitter melons should not be peeled, but the seeds and pulpy core need to be removed. Slice them thickly across or, if the melons are to be stuffed, halve them lengthwise. To tame their bitterness, sprinkle them generously with salt, leave them to stand for at least 30 minutes, then rinse well and pat dry. For a 'belt and braces' approach, blanch them in boiling water as well.

## Cooking

With the exception of bitter melons, there is no mystery to cooking exotic cucurbits. They can be stuffed, braised, steamed or stir-fried, and are particularly delicious with strongly flavoured oriental sauces. They are also excellent in spicy curries and casseroles, since their spongy mild flesh soaks up the juices and flavours of the ingredients with which they are cooked.

Try a mouthwatering dish of Curried Loofahs with Prawns (page 62), served with cooling yoghurt and crisp warm poppadoms. Alternatively, fry cubed bottle gourd with onion, chilli, ginger, garlic, a dash of turmeric, paprika, garam masala and salt. Moisten the mixture with lime juice and a little water, then cover it and simmer gently until the flesh is tender – it will cook in its own steam. For a more substantial dish, you could add some precooked yellow split peas. Finish off with a final 'tempering' of fried dried red chilli, cumin seeds, sizzling butter and a sprinkling of chopped fresh coriander.

Bitter melons are a staple in the Canton region of China, where they are usually stir-fried with salted black beans and garlic, or stuffed with prawns and served with black bean sauce. The Indians stuff whole bitter melons with onions, spices and nuts or meat, then tie them with thread before frying them in oil or simmering them in a tomato-flavoured broth. They are also a mainstay in curries, pickles and sambals.

## Nutrition

Being over 90 per cent water, most exotic cucurbits contain unspectacular levels of nutrients. The snake gourd provides a little carotene and a useful amount of magnesium, and the bottle gourd contains zinc and vitamin C. Most other vitamins and minerals are present, but in minuscule amounts.

The one exception, however, is the bitter melon, which is

positively brimming with all things good; and, of course, it would have to be the one that tastes like medicine. One small melon contains over three times more potassium than the other exotic cucurbits, three times the daily requirement of vitamin C, plenty of folate and carotene, as well as a modest amount of magnesium and iron. Although some nutrients will inevitably be lost during the salting and blanching process, the vegetable nevertheless seems to be one that we should be eating more often.

## Varieties

Compared to other vegetables, there is only a small selection of exotic cucurbits in the seed catalogues. Most are listed under generic names, though specific varieties are beginning to make an appearance.

Tasting trials were carried out in late summer. The gourds were unpeeled, thickly sliced and tested two ways: lightly steamed, and fried in a tasteless oil.

### Loofah

The elongated fruits of the loofah come in two basic models: smooth *(Luffa aegyptiaca,* syn. *L. cylindrica)* and angled or ridged *(L. acutangula)*. Also known as the sponge gourd, the smooth loofah, if allowed to age, produces a fibrous interior that is a necessary accoutrement for serious bathers. It is, however, perfectly edible when harvested young: the fruits are soft with no hint of fibre.

Two varieties we grew were the angled '**Sureka**' and the smooth-skinned '**Harita**'. Both have a lovely spongy texture, crisp skin and a unique aroma – something like warm nuts or toast. Unfortunately, some of our fruits suffered from a touch of bitterness, which – while desirable in a bitter melon – was not welcome here. The best samples, however, had a nice earthy flavour.

### Tinda

Tinda *(Praecitrullus fistulosus)* is an odd but versatile little vegetable, grown primarily in India. Seed is sold under the uninspiring name of **tinda**. Our tastings revealed a mild, pleasant but hard to describe flavour – toasty, sweet, nutty perhaps. They are delicious served whole and stuffed with fried onion, garlic and tomatoes, or with minced lamb, onion and curry spices.

**Snake gourd**

With severely elongated striped fruit that can reach lengths of 1.5 metres/5 feet or more, the snake gourd *(Trichosanthes cucumerina)* is well served by its name. Carrying one home through a crowded street can be a little awkward, but its distinctive flavour and texture make the effort worthwhile.

The seeds we bought were simply called **snake gourd**. Disappointingly, the fruits failed to reach snake-like dimensions, stopping at about 45 cm/18 inches in length. Even so, they were among the best-flavoured gourds and definitely worth growing. The flesh was beautifully sweet, vaguely reminiscent of French beans, and with no trace of bitterness. The texture was dense and silky, delicious to bite into and perfect for stir-fries, soups and casseroles.

**Fuzzy gourd**

The fuzzy gourd *(Benincasa hispida)* is a squat green fruit with barely discernible spots and a five o'clock shadow. We tried one called **Chinese green fuzzy gourd** and liked the crisp, juicy flesh, although it was almost devoid of flavour. Cut into small cubes and cooked in proper home-made stock, however, it makes a wonderfully comforting and quickly prepared oriental-style soup (page 62).

**Bottle gourd**

The bottle gourd *(Lagenaria siceraria)*, or dudhi, shows wide variation in form and function. The hard shells of some varieties are used as bowls and bottles in the tropics, suggesting the possible origin of the vegetable's English name. A number of other varieties have been selected for their culinary qualities, and they come in all manner of shapes and sizes, ranging from small round specimens to those shaped like a large club.

We grew a roundish Indian variety sold under the unlikely name of '**Dudhi (Bottle Gourd MGH–1)**'. It had a pleasant, though mild, flavour. We also liked the long thin fruit of the Italian **cucuzzi** or **longissima**. Picked when 2.5–4 cm/1–1½ inches in diameter, it had nice firm-textured flesh and a sweet flavour.

### Bitter melon

With a light to dark green skin covered with warts, the bitter melon *(Momordica charantia)* is an oddly attractive vegetable, though westerners should approach it with caution: the bitter flesh may take some getting used to. A named variety we grew was '**High Moon**'. It has greenish-white skin, large rounded warts and a crisp texture. We had a problem with the bitterness at first, but found it could be modified by the salting and blanching technique (see Preparation, page 57). We also discovered that stuffing the melons with a sweetish spicy mixture of pork, coconut, sugar, fennel and coriander (below) tamed the bitterness. Once we had survived the first timorous bite, we kept coming back for more.

## Pan-Fried Bitter Melons with Pork, Coconut & Coriander Stuffing

**Serves 4 as a light meal with rice**

*4 small bitter melons, about 13 cm/5 inches long, unpeeled*
*2 tablespoons salt*
*2 teaspoons sugar*
*40 g/1½ oz dried coconut flakes*
*1 teaspoon muscovado sugar*
*½ teaspoon fennel seeds*
*½ teaspoon cumin seeds*
*½ teaspoon coriander seeds*
*groundnut oil, for frying*
*2 shallots, thinly sliced*
*2 garlic cloves, thinly sliced*
*140 g/5 oz minced pork*
*1–2 teaspoons dried chilli flakes*
*2 tablespoons chopped fresh coriander*
*salt and freshly ground black pepper*
*fresh coriander sprigs and lime wedges, to garnish*

**If you eat a bitter melon only once, try it this way. The sweetish flavours of the stuffing — a mixture of pork, fennel, coconut, chilli and fresh coriander — stand up to the quinine-like bitterness of the melon. The combination is surprisingly pleasant.**

**1** Slit the melons in half lengthwise. Scoop out the pulp and seeds, and discard. Rub each half inside and out with the salt and sugar. Leave to sit, cut side down, for 30 minutes.

**2** Bring a large saucepan of water to the boil. Meanwhile, rinse the melons under cold running water, then blanch them for 10 minutes in the boiling water. Drain them and plunge into cold water. Leave to cool, then drain and pat dry thoroughly.

**3** Preheat the oven to 200°C/400°F/gas 6. Spread the coconut flakes on a baking tray and toast in the oven for 2–3 minutes, until beginning to turn golden. Do not let them burn.

**4** Set aside one-third of the flakes as a garnish and put the rest in a blender with the sugar, fennel, cumin and coriander seeds. Process to a coarse powder.

**5** Heat 2 tablespoons of oil in a wok or frying pan and stir-fry the shallots and garlic over medium heat for 1 minute. Stir in the meat and the powdered coconut mixture. Add the chilli flakes and fresh coriander, and season with salt and pepper. Continue to stir-fry until the meat is no longer pink. Allow to cool a little, then use the mixture to stuff the melon halves.

**6** Choose a frying pan big enough to take the melon halves in a single layer and pour in enough oil to come 1 cm/½ inch up its sides. When the oil is hot but not smoking, carefully add the melon halves, cut side up. Reduce the heat slightly, and fry for about 3 minutes, until the underside is golden and the stuffing is heated through.

**7** Remove with tongs and drain on paper towels. Transfer to a warmed serving dish. Sprinkle with the reserved toasted coconut, garnish with coriander sprigs and lime wedges and serve.

## Curried Loofahs with Prawns▸

**Serves 4–6 as a light meal**

*2 teaspoons coriander seeds*
*1 teaspoon cumin seeds*
*4 small angled loofahs, weighing about 675 g / 1½ lb*
*3 tablespoons groundnut oil*
*350 g/12 oz raw shelled tiger prawns*
*½ teaspoon mustard seeds*
*½ teaspoon asafoetida (optional)*
*¼ teaspoon freshly ground black pepper*
*¼ teaspoon garam masala*
*1 teaspoon salt*
*½ teaspoon ground turmeric*
*1 onion, halved and thinly sliced*
*2–3 small red chillies, deseeded and sliced*
*4–6 tablespoons water or stock*
*2 plum tomatoes, peeled, deseeded and chopped*
*chopped fresh coriander, to garnish*

**The spongy flesh of the loofahs soaks up all the flavoursome juices and becomes meltingly soft. You can use courgettes instead, but the flavour and texture will not be quite the same. Serve with a bowl of yoghurt and some poppadoms or warm naan bread**

**1** Put the coriander and cumin seeds in a small heavy-based pan without any oil. Dry-fry over moderate to high heat until they smell fragrant, taking care not to let them burn. Remove from the pan and grind to a coarse powder with a pestle and mortar.

**2** Using a swivel peeler, shave away the edges of the loofahs' sharp ridges. Cut the flesh at an angle into 1 cm/½ inch slices.

**3** Heat the oil in a large frying pan until almost smoking. Add the prawns and fry for a few minutes until crisp. Remove with a slotted spoon and drain on paper towels.

**4** Reduce the heat to medium and add the mustard seeds and the asafoetida, if you are using it. When the mustard seeds begin to splutter, stir in the ground cumin and coriander seeds, black pepper, garam masala, salt and turmeric. Stir for 20 seconds, then add the onion and cook for 5 minutes, until it has softened. Next add the chillies and cook for 2 minutes more.

**5** When the onion and chillies are soft, stir in the loofah slices with enough water or stock to moisten. Cover and cook for 5 minutes, stirring occasionally, until almost tender.

**6** Add the tomatoes and prawns and cook for another 5 minutes. Sprinkle with coriander just before serving.

## Fuzzy Gourd Soup

**Serves 4**

*2 teaspoons groundnut oil*
*2 shallots, thinly sliced*
*450 g/1 lb fuzzy gourd, peeled and cut into 2 cm/¾ inch pieces*
*8 small shiitake mushrooms, diced*
*115 g/4 oz minced pork (optional)*
*700 ml/1¼ pints Chinese stock (page 98) or home-made chicken stock*
*½ teaspoon shoyu*
*½ teaspoon sancho (Japanese pepper)*
*½ teaspoon salt*

**If you can't find any fuzzy gourds, use four peeled courgettes or a marrow instead. Home-made stock is essential. Replace the Japanese pepper with crushed Szechuan peppercorns, or even ordinary black peppercorns, if necessary.**

**1** Heat the oil with the shallots over moderate heat and fry for 30 seconds, until the shallots are soft but not coloured.

**2** Add the fuzzy gourd, mushrooms and pork (if using). Stir quickly, then pour in the stock and add the shoyu, sancho and salt. Bring to the boil, then simmer for 20 minutes or until the melon is tender.

# CUCUMBERS

The cucumber *(Cucumis sativus)* was highly esteemed by the Victorians, who gave it favoured status at their tables. Given the Victorian penchant for propriety, though, the cucumber would probably have been banned had its exotic sex-life been exposed.

Normal reproduction in the cucumber pretty much follows the usual procedure: the sexes are distinct and each plant has separate male and female flowers. Insects help move pollen from male to female flowers, which then develop into seed-containing fruits. This is only part of the story, however, and some cucumber varieties deviate from the usual male-female approach.

One of the more common variations on reproduction is known as parthenocarpy, which simply means that female flowers spontaneously produce fruits without pollination. Because there is no pollination, the resulting fruits are seedless. There are also varieties that develop female flowers only. Plant breeders have taken advantage of this tendency by combining it with parthenocarpy to produce greenhouse plants that give us the long seedless fruit commonly found in the shops. Do not mistake the small jelly-like structure in the middle of the fruit for seeds. They are not seeds, but would have become so had pollination occurred.

## A bitter pill

Older cooks and gardeners may still have vivid memories of an unpleasant bitter taste interrupting their enjoyment of a cool, refreshing cucumber. This bitterness is caused by chemicals called cucurbitacins, the most bitter substance known – they can be tasted in concentrations as low as one part per billion. Research carried out in the 1950s identified a gene controlling their production. Fortunately many modern varieties now have the bitterness bred out of them.

## Cultivation and harvesting

The best cucumbers for growing in greenhouses and tunnels are the all-female parthenocarpic varieties. Under certain growing conditions, however, male flowers may be produced. In such cases, male flowers should be removed immediately, as misshapen fruit containing seeds will develop if pollination occurs.

Some outdoor varieties are also parthenocarpic. This means that seedless fruits can be produced when the plants are grown under some sort of fleece to bring on earlier production. If pollination does occur, well-shaped fruits will still develop, although they will have seeds.

Outdoor cucumbers must be picked when they are still young, since the seeds become tough once they exceed a length of 5 mm/¼ inch. The skin also toughens up as the fruit matures, and peeling becomes a necessity.

Timing of the harvest is less critical with parthenocarpic greenhouse cucumbers, as these are seedless and have a thinner skin than outdoor varieties. We have picked some real overgrown monsters, but the fruits were still tender.

## Buying and storing

When selecting a cucumber, what you feel is more important than what you see. Run your hands over the entire length – the cucumber should be firm and stiff rather than feeble and flabby. Cucumbers do not like intense cold and are best stored in a cool larder or the salad drawer of the fridge. Once you've cut a cucumber, keep it tightly wrapped in cling-film, including the cut end, since exposure to the air will hasten deterioration.

'Bedfordshire Prize Ridge'

'Vert de Massy'
'Bianco Lungo di Parigi'
'Crystal Apple'
West Indian gherkin
'Aria'
'Jazzer'
'Carmen'
'H-19'

## Preparation

If the qualities of the cucumber are to be fully appreciated, it is best to eat it raw, unsalted and preferably cut into slices thick enough to get your teeth into. Baking or frying cucumber is simply missing the point.

If you must cook your cucumber, or if its sprightliness needs taming for a particular type of salad, then it's better to remove some of the moisture first. The quickest and least aggressive way to do this is to sprinkle the cucumber – sliced, cubed or whatever – with a pinch of salt, a little sugar and a teaspoonful or so of wine vinegar or lemon juice. After just a few minutes, liquid will pour forth and the cucumber will be left with much of its flavour and texture intact.

Smooth-skinned greenhouse cucumbers have a tender skin and usually do not need peeling. Indeed, the skin adds flavour, as well as colour and texture, to the finished dish Outdoor varieties, however, have thicker skins that may be tough and need peeling.

Some people complain that cucumber skin gives them indigestion, others claim that it prevents it. If digestive issues are a concern, the so-called 'burpless' varieties may be an answer. Alternatively, rather than entirely denude a cucumber of its skin, a reasonable compromise is to remove alternate strips along the entire length.

Cucumbers can be fashioned into any number of pleasing shapes according to whim or the nature of the dish. Use a mandoline to cut them into paper-thin slices, or cut semi-peeled or grooved cucumbers into thick diagonal slices. To make matchsticks, stack a pile of unpeeled slices and cut them across. For crescents, halve the cucumber lengthwise, scoop out the seeds with a coffee spoon, and slice across; or make 'boats' for stuffing by cutting the halved and deseeded cucumber into short lengths. Alternatively, slice a whole cucumber across into elongated chunks, then hollow out the centres. Stand the chunks upright and stuff with herbed cream cheese, prawns, or sushi-style rice.

## Cooking

Like bananas and oranges, cucumbers are with us all year round – dependable, familiar and understated. As such they fail to get the rave reviews allocated to more flamboyant species. In fact, cucumbers are one of the most utilitarian and versatile of vegetables and offer numerous possibilities to the cook.

We could usefully take a lead from oriental cooks who understand how well cucumbers pair with their opposites – a case of the yin and yang at work in the kitchen. For example, their mild, slightly sweet flavour and crisp texture counteracts pungent, sour or soft ingredients.

Their cooling nature is exploited to the full in the cuisines of the Middle East and India, where cucumbers are traditionally combined with yoghurt and mint to make a soothing dip or sauce to serve with grilled meat or fish, or to offset the heat of a spicy curry. In Greece, they are cut into chunks and served in appetising salads with tomatoes, crisp lettuce, olives, feta cheese and mint.

Although mint is the herb most often associated with cucumber, it's worth remembering that salad burnet and borage flowers make stunningly beautiful and pleasing additions to cucumber-based salads. Both taste faintly of cucumber and work synergistically to create a more positive flavour.

Cucumbers are almost synonymous with East European cuisine. Here, a traditional treatment is to slice them thinly, salt them, sprinkle with dill and serve as a refreshing accompaniment to fatty meat dishes. They are also the mainstay of pickles and, for some people, few gustatory experiences would be complete without them. To the uninitiated, pickle varieties are confusing and seemingly not very different: sour, half-sour, sweet-and-sour, dill, new green (freshly made) and haimishe (home-made). To the enthusiast, however, the appropriate choice of pickle is crucial. We are very fond of the good old-fashioned bread and butter pickle: to make it, layer 1.3 kg /3 lb thickly sliced cucumbers with a couple of sliced onions and peppers, and a small sliced chilli. Add salt and leave overnight to drain. Put 350 g/12 oz sugar in a saucepan with 850 ml/1½ pints of white wine vinegar, 1 tablespoon each of mustard seeds and dill seeds and 1 teaspoon of ground turmeric. Boil the sugar and vinegar mixture for 5 minutes, then add the vegetables. Cook for 5–10 minutes, depending on how crunchy you like your pickle. Leave the pickle to cool, pour it into clean jars and store it in a cool place.

Pickled cucumbers are a feature of oriental cuisine, too, although these delicate versions are far removed from their rough-

and-ready East European cousins. For Japanese cucumber and sesame pickle, peel and slice small pickling cucumbers wafer-thin, sprinkle them with a dash of salt, sugar and rice vinegar, then squeeze them dry and mix with shoyu, crushed toasted sesame seeds, more rice vinegar and perhaps some pink pickled ginger. Served chilled, this makes an impeccable counterpart to white rice or noodles.

The Chinese use cucumbers in stir-fries, deftly cutting them into matchsticks to sizzle with shredded chicken and sometimes a mustardy sauce. Cucumbers also feature in Thai and Indonesian cooking, adding texture to stir-fries and soothing the heat of chillies in freshly made pickles.

Cucumbers go well in salads with crisp-textured vegetables, such as radishes, fennel or jicama, as in the Mexican salad Pico de Gallo (page 168

. Alternatively, you can try a fiery but cooling cucumber and mango pickle: mix peeled and finely diced cucumber with diced mango, finely chopped fresh green chilli, fresh coriander, lime juice, salt and sugar. Then sizzle some mustard seeds and a pinch of asafoetida, if you have it, in a tablespoon of hot oil and pour over the pickle while still hot.

## Nutrition

Being over 95 per cent water, the cucumber contains extremely low levels of all nutrients, but it does contain a small amount of vitamin C. If the skin is left intact, the amount of carotene present is significantly increased.

## Varieties

Cucumbers provide an abundance of choice for the gardener. There is a minefield of classification systems, based on growing techniques, country of origin, physical characteristics, reproduction method and use, so it is best simply to pick up a seed catalogue and try a couple of varieties that tempt you. We grew only the trailing types rather than bush varieties.

Outdoor varieties include '**Vert de Massy**', a French cornichon pickling type used when they are the size of a little finger or smaller. Though they may be good for pickling, they are definitely not for eating raw. An American parthenocarpic variety is '**H-19**'. Fruits are typically short and fat, with spines and warts scattered over their light green skin. Though they are recommended for pickling, they are also delicious when eaten fresh.

The old English variety '**Bedfordshire Prize Ridge**' has green skin with large warts and black spines. When eaten young, the fruits have a good texture and flavour, though some of them can be bitter. A modern parthenocarpic hybrid is '**Jazzer**'. The skin is a deep green with black spines, while the fruits have a crisp texture and rich cucumber flavour.

'Burpless Tasty Green' (not shown) lives up to its name – it tastes delicious and does not cause digestive problems. The plants produce long fruits and need trellising if the fruits are to grow straight.

The oddly shaped '**Crystal Apple**' is somewhat round, with a large seed cavity and ivory-coloured skin that turns yellow when mature. Picked while still small, the fruits are surprisingly delicious, although they can occasionally be bitter.

'**Bianco Lungo di Parigi**' is a small, whitish-skinned variety with a reasonable flavour, though the fruits are sometimes bitter. The **West Indian gherkin** (*Cucumis anguria*), which belongs to a different species, has small green oval fruits, covered with soft spines that give 'texture' a new meaning. Appearances can be deceiving, though, and the young fruits have both a good texture and an excellent flavour. Use them in spicy relishes and oriental pickles. They need the protection of a greenhouse or polytunnel if they are to do well in a temperate climate.

'**Carmen**' is a typical modern greenhouse cucumber. Green and seedless with spineless tender skin, it is a highly productive variety well worth growing. It is superb in Cucumber and Lemon Thyme Ice-Cream(page 68), and is also good combined with smoked fish in a cucumber and smoked fish pâté. For two servings, coarsely grate about 115 g/4oz unpeeled cucumber and mix with 85 g/3 oz each of ricotta cheese and flaked smoked mackerel or trout, some dill seeds or chopped fresh dill, a squeeze of lemon juice, and salt and pepper. Serve well chilled, with dark rye bread or triangles of hot toast.

If we were to grow only one type of cucumber, however, it would be the small-fruited '**Aria**'. It is seedless with an excellent flavour, tender skin and crisp-textured flesh.

## Cucumber, Sea Vegetables & Chargrilled Squid▸

**Serves 4 as a starter**

*15 g/½ oz dried arame, soaked in cold water for 5 minutes*
*4–6 shiitake mushrooms, thinly sliced*
*3 tablespoons extra-virgin sunflower oil*
*600 g/1 lb 5 oz squid, quills and tentacles removed*
*2 tablespoons shoyu*
*3 small tender-skinned seedless cucumbers*
*a few slices of Japanese pink pickled ginger (optional), to garnish*

**For the wasabi dressing:**

*5 tablespoons rice vinegar or cider vinegar*
*1 tablespoon sugar*
*1 teaspoon shoyu*
*¼ teaspoon salt*
*¼ teaspoon wasabi paste*

**For this dish, you will need a seedless cucumber such as 'Carmen' or 'Aria'.**

**1** Drain the arame, put it in a saucepan with fresh water just to cover and simmer for 20 minutes. While the arame is cooking, gently fry the mushrooms in 1 tablespoon of the oil. Drain on paper towels and leave to cool. Combine the dressing ingredients.

**2** Slit the squid pouches down one side and open them out flat. Trim the edges to form a triangle. Rinse under cold water and pat dry. Lightly brush with oil.

**3** Drain the arame, reserving the cooking water. Add to the pan with the shoyu and remaining oil. Stir over medium-high heat for a minute, then add 4 tablespoons of the cooking water and cook until the liquid has almost evaporated. Spread out on a plate and leave to cool.

**4** Shave away a broad band of skin from opposite sides of each cucumber. Then cut the cucumber lengthwise into very thin slices. Stack the slices and cut them diagonally into 4 cm/1½ inch pieces. Put in a bowl with the arame and mushrooms, and toss with the dressing.

**5** Heat a clean ridged stove-top grill pan. When it's very hot, place the squid pieces on it and grill until they start to curl, about 30 seconds. Turn them over and grill for a few more seconds.

**6** Divide the cucumber mixture between four plates and add the squid. Spoon over any dressing left in the bowl and garnish with a few slivers of pickled ginger, if you are using it.

## Cucumber & Lemon Thyme Ice-Cream with Mixed Leaf Salad

**Serves 4 as a starter**

*500 g/1 lb 2 oz tender-skinned cucumber*
*125 ml/4 fl oz stock*
*250 g/9 oz ricotta cheese*
*2 tablespoons lemon juice*
*2 tablespoons chopped fresh lemon thyme*
*1 teaspoon sea salt*
*¼ teaspoon freshly ground black pepper*
*4 good handfuls of mixed robust salad leaves*
*6 small radishes, thinly sliced*
*thin cucumber sticks, to garnish*

**For the lemon vinaigrette:**

*1 tablespoon lemon juice*
*1 teaspoon finely chopped fresh lemon thyme*
*½ teaspoon Dijon mustard*
*6 tablespoons extra-virgin sunflower oil*

**For the salad, use a mixture of leaves such as rocket, red chicory, purslane and lamb's lettuce. Ice-cream and salad may sound a strange combination, but the contrast of texture and temperature is mouthwatering. You'll want to lick your plate afterwards, as the ice-cream melts into the dressing.**

**1** Coarsely chop the cucumber. Put all the ingredients except the salad leaves, the radishes and the cucumber garnish in a blender and process until smooth. Either freeze in an ice-cream maker or pour into a shallow container, cover with cling-film and freeze for about 2 hours or until beginning to harden round the edges. Tip into a bowl, whisk until smooth and freeze again. Repeat the process once more if you have time. About 30 minutes before serving, put in the fridge to soften.

**2** Arrange the leaves and the radishes on individual plates. Whisk the dressing ingredients together with seasoning to taste and spoon over the leaves. Place a scoop of ice-cream in the centre and garnish with the cucumber sticks.

# GLOBE ARTICHOKES & CARDOONS

Globe artichokes *(Cynara cardunculus* Scolymus Group*)* and cardoons *(C. cardunculus)* share the ability to polarize gardeners and cooks: either you love them or you hate them. Apart from this divisive tendency, however, they appear to have little in common, from either a horticultural or a culinary perspective.

Globe artichokes, for example, are short-lived perennials best propagated by shoots taken from existing plants, whereas cardoons are treated as annuals and started from seed each year. In addition, we eat the flower buds of globe artichokes, and can only wonder in awe at the courage, or possibly desperation, of the first person to try them. In the case of cardoons, though, we eat the thick fleshy leaf stems.

Things are not always as they seem, however, and these supposedly disparate vegetables are, in fact, both members of the thistle family. Scientists have a theory that cultivated cardoons were selected in antiquity from wild forms growing in the Mediterranean region, and that the first globe artichokes probably descended from cardoons with the biggest and best flower buds.

We tested this theory, allowing some of our cardoons to grow on unharvested. Eventually they produced flower buds, albeit spikier-looking and harder than those of globe artichokes. Other than that, the resemblance was startling. The cooked cardoon buds also had a strong artichoke-like flavour, although they were more troublesome than artichokes to prepare and eat.

## Cultivation and harvesting

Globe artichokes and cardoons are big, bold, architectural plants, sometimes reaching 1.4–1.8 m/4–6 feet in height. As such, they suit the ornamental border as well as meriting space – and they will need plenty of it – in the vegetable plot.

The artichoke globe is made up of a hairy central choke surrounded by spine-tipped bracts or leaves. Harvesting should take place while the bracts are still tightly closed, cutting off the globe with a short length of stem attached. If harvesting is delayed, the choke – which is made up of tiny florets – matures and becomes larger, transforming the globe into a stunning, but inedible, purple-blue thistle.

Cardoons are normally blanched in late summer or early autumn to reduce bitterness and toughness. Although a number of blanching techniques have been devised, it is probably just as effective simply to wrap up the huge stems in sheets of black plastic and tie these in place with raffia or string. Harvesting can then take place 4–5 weeks later.

## Buying and storing

Choose globe artichokes with firm, tightly packed heads and a vibrant fresh colour. Reject any with withered brown or open leaves, and also check for brown spots at the base of the head, as this could indicate the start of decay.

With cardoons, look for those with sprightly leaves and firm, plump ribs that do not feel hollow when tapped. There should be no bruising or brown patches. In Britain, cardoons tend to be elusive, but specialist food shops and upmarket greengrocers sometimes stock them in late autumn and winter.

Both vegetables should be eaten as fresh as possible, preferably on the day of purchase. If you have to store them, use damp newspaper to wrap cardoons or artichokes sold with long stems. Keep in a cool moist place for no more than two days. Large single globe artichoke heads can be tightly wrapped in cling-film and stored in the fridge for two or three days.

Boiled artichokes can be covered with cling-film and stored in the fridge for a day or two, but after that they may develop an off-putting verdigris colour and are probably best discarded.

## Preparation

To prepare globe artichokes for boiling or steaming, bend back the outer leaves, snapping them off just above the fleshy part, or cut off the tips with scissors. Keep cutting or snapping off leaves above the fleshy part until you are left with a central cone of paler thinner leaves. Using a large sharp knife, slice about 2.5cm/1 inch from the top of the cone parallel with the base. Rub all the cut surfaces with a piece of cut lemon to prevent them from turning black. It is better to remove the hairy choke after cooking, when it can more easily be spooned out.

'Green Globe'
Violetta
GLOBE
ARTICHOKES

To prepare baby artichokes for frying whole, pull off and discard 10–12 of the outer leaves, then slice off the top 2.5 cm/ 1 inch of the remaining leaves. If the choke is very small and tender, it need not be removed; otherwise, force a pointed coffee spoon into the centre, rotate it and scoop out the barely formed innermost leaves and the choke.

Cardoons are easy to cook but have a reputation for being tiresome to prepare. In fact, they are no more trouble than a head of celery. First, discard the base and the tough outer stems. Separate the remaining stems, cutting off the foliage and any prickles, and wash them. Cut the heart lengthwise into quarters and slice the stems into 7.5 cm/3 inch pieces, rubbing all cut surfaces with a piece of lemon to prevent discoloration. Then bring to the boil a large pan of water salted and acidulated with lemon juice, drop in the stems and hearts, boil them for 15–30 minutes (depending on the amount of cooking to follow), and drain. Some recipes specify boiling for up to 3 hours, but this seems to be nonsensical.

When they are cool enough to handle, lift the edge of the skin on the underside of each rib with your fingernail and peel it away. With a bit of luck it will come away in wide strips. Next remove the fibrous strings. Since these go all the way through the rib, it is a matter of judgment as to when to stop – you could be left with nothing. You can usually get by with peeling away enough to smooth out the striations on the top side of the rib, then taste and see. If the remaining strings are tough, peel away a few more. The cardoons can now be submerged in acidulated water for several hours if necessary, before you proceed with your chosen recipe.

## Cooking

Globe artichokes can be cooked and served in a number of ways, depending on their size. To cook large ones, bring a big pan of acidulated salted water to the boil, drop in the prepared artichokes and weigh them down with a heatproof plate so they remain submerged. Simmer them briskly for 30–40 minutes, until they are tender and an outer leaf pulls out easily. Drain and leave them to stand upside down to allow excess water to run off. Artichokes can also be steamed, and we found that this produced a more intense flavour.

Large globes can be stuffed with a seafood mixture – anchovies, parsley and tomatoes, for instance – or mushrooms, garlic and herbed breadcrumbs. Alternatively, the leaves can be detached one by one and dipped into an appropriate sauce, perhaps a lemony vinaigrette or home-made mayonnaise if you are serving the vegetable cold, hollandaise sauce or simply melted butter if serving it hot. Suitably anointed, each leaf is passed decorously between the teeth and the fleshy portion scraped off. The best is yet to come, however, for once all the leaves have been dealt with there remains the *fond* or bottom, a delicacy in its own right, to be cut up and eaten with a knife and fork.

Small artichokes are tender enough to be deep-fried, either whole or cut in wedges. The outer leaves become beautifully crisp, enclosing a soft tasty heart. Plunge them into very hot oil and fry them for about 5 minutes, covering the pan so that they cook in a mixture of steam and oil. Once they are cooked, sprinkle them with lemon juice, some sea salt flakes, coarsely ground black pepper and a showering of chopped flat-leaved parsley, and eat them at once.

Small globes can also be braised in oil and wine, perhaps flavoured with mint. Force about a teaspoonful of chopped mint, garlic and lemon zest into the centre of each prepared globe and, if you can, between the leaves. Braise them, covered, in a mixture of olive oil, white wine and water until they are tender, seasoning to taste. Remove, allow to cool, then sprinkle with more of the mint mixture and season as necessary. Boil down the juices to use as a dressing.

Cardoons are generally eaten plainly boiled, either in water or stock, and served with a béchamel sauce, sometimes flavoured with onion or ham. The Spanish like to add almonds or walnuts. Cardoons are also delicious in meat or vegetable stews, and they add a distinctive flavour to soups. They can be used in any recipe calling for celery or fennel.

Try them with roasted peppers and anchovies in a richly flavoured gratin (page 74). Alternatively, make a salad of cooked ribs, baby broad beans and olives, tossed in a lemony dressing and garnished with plenty of mint or dill.

Early in the season, the tiny leaves around the heart can be eaten raw, either simply dressed with lemon juice and olive oil or included in a selection of vegetables for dipping in *bagna cauda* – a hot soup-like dip from Piedmont in Italy, based on anchovies and olive oil.

## Nutrition

Both globe artichokes and cardoons contain small amounts of protein, folate, potassium, calcium and magnesium.

## Varieties

Since we do not grow globe artichokes ourselves, we tried samples provided by friends, specialist growers and importers. The names given below are not necessarily the correct variety names, but they are the names under which the varieties we tried were sold.

Our favourite, which goes by the name of **Spinosi** (not shown), is sold on a long stem while the buds are small. The beautiful, but viciously thorny, lime-green leaves are elegantly tipped with gold. The flavour is rich and meaty, with a background sweetness slightly reminiscent of parsnips. Once divested of thorns, the entire head can be eaten, rather than just the fleshy leaf base. The choke, which has shorter hairs than most, is easily removed.

Another small-headed type, sold as **Violetta**, has handsome purple-green elongated globes with unobtrusive chokes. It has an assertive artichoke flavour, although it is not as sweet as Spinosi. Once cooked, the heads are tender enough to eat whole, and make an excellent salad with rocket and olives (page 74).

Of the large-headed types, '**Green Globe**' is the most readily available. It produces reasonably flavoured pale green roundish heads with succulent, fleshy-based leaves that are ideal for stuffing or dipping in a sauce.

Because of the size of the plant we grew only one variety of cardoon, '**Gigante di Romagna**'. It produced thick, solid thorny ribs, which became reasonably tender without prolonged cooking. It had a good assertive artichoke-like flavour, slightly astringent, and when sampled in late summer was not especially bitter or tough, even though we did not blanch the leaves before harvesting. Later in the year the ribs became decidedly bitter, so we recommend wrapping your cardoons in black plastic (see Cultivation and harvesting, page 70) if you are going to harvest them in autumn or winter.

## Baby Artichoke, Rocket & Olive Salad▸

**Serves 4**

*4 small artichokes, such as Violetta, about 6–8 cm/2½–3 inches in diameter*
*juice of ½ lemon*
*a good handful of rocket*
*4–5 Kalamata olives, pitted and halved*
*1½ tablespoons extra-virgin olive oil*
*sea salt flakes*
*cracked black pepper*
*cayenne pepper (optional)*
*a few Parmesan shavings*

**As its name suggests, the small Violetta artichoke is deeply tinged with purple. It looks beautiful with purple-black Kalamata olives.**

1 Put a large pan of water acidulated with the lemon juice to heat.

2 Trim the artichoke stalks at the base. Pull off 10–12 of the outer leaves, until the visible leaves are mostly pale green and purple. Using a sharp knife, slice off the top 2.5 cm/1 inch of the remaining leaves. Rub the cut surfaces with the squeezed lemon half to prevent blackening.

3 When the water is boiling, plunge the artichokes into it. Cook for about 35 minutes, until just tender, then drain.

4 When the artichokes are cool enough to handle, slice them in half lengthwise. Scoop out the small hairy choke.

5 Arrange some rocket leaves on a serving plate and place the artichoke halves on top, cut side upwards. Add the olives and trickle over the olive oil. Sprinkle with sea salt flakes, cracked black pepper, a light dusting of cayenne if you like, and a few shavings of Parmesan.

## Cardoon, Roasted Pepper & Anchovy Gratin

**Serves 2–4**

*450 g/1 lb cooked cardoon stems, strings and skin removed (see Preparation, page 72)*
*6 tablespoons olive oil, plus more for the dish*
*1 small roasted red pepper, peeled, deseeded and cut into small squares*
*4–6 canned anchovy fillets, drained and chopped*
*salt and freshly ground black pepper*
*85 g/3 oz breadcrumbs, made from a slightly stale loaf*
*finely grated zest of 1 lemon*
*3 tablespoons chopped fresh flat-leaved parsley*
*4 tablespoons freshly grated Parmesan cheese*

**You will need a cardoon head weighing about 1kg/2¼ lb when trimmed of its base and coarse outer leaves (this preliminary trimming will have been done if you are buying your cardoon). This will give you about 450 g/1 lb of stems after a final trimming of foliage and spines.**

1 Preheat the oven to 200°C/400°F/gas 6. Slice the cooked cardoon stems into 2.5cm/1 inch pieces. (If any of the remaining fibrous strings still seem tough, here is a second opportunity to remove them.)

2 Place the stems in a lightly oiled shallow baking dish. Scatter over the red pepper and anchovies, and season with salt and pepper, bearing in mind the saltiness of the anchovies.

3 Combine the breadcrumbs, lemon zest and parsley with a few more grindings of black pepper. Bind the mixture together with 4 tablespoons of the olive oil and scatter this over the cardoons. Sprinkle with the Parmesan and the remaining olive oil. Bake at for 20 minutes, until the top is golden and crisp.

Broccoli raab
'Broccoletto'
Broccoli
'Late Purple Sprouting'

# BROCCOLI & CAULIFLOWER

Cooks, gardeners and supermarkets are often at odds when it comes to naming various types of broccoli and cauliflower. Taxonomists, however, classify cauliflower as *Brassica oleracea* Botrytis Group and broccoli as *B. oleracea* Italica Group.

A distinguishing feature of cauliflower is a head consisting of a dense 'curd', usually white but sometimes orange or even green. No flower buds are evident, though flowers would eventually be produced if the heads were left uncut on the plant. Using the curd criterion, the lime-green 'Romanesco', though often called a broccoli, is actually a type of cauliflower; as also, even more confusingly, are some varieties of late white sprouting broccoli.

In the case of broccoli, the head is made up of tightly packed 'beads', each one of which is a tiny flower bud. If you look closely enough, you will see that this characteristic is typical of early and late purple sprouting broccoli, early white sprouting broccoli and the familiar densely packed green heads of calabrese, called 'broccoli' in the supermarkets. Overwintered purple Cape broccoli also falls into this group.

In the case of broccoli raab, also known as rapini or *cime di rapa* (Italian for 'turnip tops'), the scientific name is *B. rapa* Ruvo Group. Evidently, despite its strong resemblance to broccoli, the taxonomists – and the Italians – feel it is more closely related to the turnip.

Cauliflower 'Snowbred'

Broccoli 'White Star'

## Cultivation and harvesting

Cauliflowers can be grown virtually throughout the year. However, there seems little point in adding them to the long list of summer vegetables; it makes more sense to grow them for harvesting in winter or early spring, when other vegetables are thin on the ground. Unfortunately, overwintering cauliflowers demand a long-term commitment, since seeds must be sown in the summer and the plants tended for months before they are ready for harvesting.

With the life-cycle of purple and white sprouting broccolis the grower embarks on a roller-coaster of emotions: enthusiastic anticipation in the summer when the plants are started off; impatience half-way though the winter; and exhilaration with the first harvests of spring. Since there is little else in the garden at this time of year, it is a crop that is really appreciated.

Sprouting broccoli gives compound returns for your investment. After the main head is cut, side-branches quickly develop to produce a second crop. As these are harvested, still more side-branches develop, though they inevitably become smaller and more laborious to pick as time wears on.

The race horses of the group are calabrese and broccoli raab. Calabrese can be harvested as soon as 12–13 weeks after sowing; broccoli raab can be ready even sooner and is good as a catch crop. Both produce side-shoots once the central head is cut.

## Buying and storing

When buying cauliflower and broccoli, freshness is all. There should be no hint of flabbiness; curds and buds should be good and solid, stems sprightly, leaves green and squeaking with life. Firmly reject any broccoli, sprouting or otherwise, with even the slightest hint of yellowing. The same applies to cauliflowers with brown spots. Choose cauliflowers with plenty of leaves enclosing the curds – they help keep the plant fresh and firm.

Loosely wrapped and stored in the fridge, calabrese will keep for a day or two but, after that, it will turn yellow and is not fit for eating. Sprouting broccoli, cauliflower and broccoli raab will keep for 4–5 days.

## Preparation

When preparing calabrese and cauliflower, remember that the stems are as tasty and edible as the heads, and, since they make up a good proportion of the vegetable – at least 50 per cent in the case of calabrese –it is a pity to discard them. In fact, a childhood treat was a raw nugget from the creamy inner part of the cauliflower's central stem; we still like to use this delicacy finely diced in salads. If you did not know what you were eating, it could easily pass for a piece of mild cheese.

The problem is that stems take longer to cook than flower heads. With calabrese, this may be overcome to a certain extent by peeling the thick central stalk, if necessary, and slicing it as thinly as possible. The slices can be used as a vegetable in their own right – they look stunning in a stir-fry with diced red pepper, for example. Otherwise they can be cooked with the florets. We leave about 2.5 cm/1 inch of stem attached to the florets, and, if they are thick, slice the flower head and stalk lengthwise into two or three pieces. This way, everything cooks in more or less the same time.

If you need to cook a calabrese or cauliflower in one piece, cut off the central stalk as close to the head as possible (save the calabrese stalk for the previously mentioned stir-fry), then make two crossed cuts in the base to allow the heat to penetrate. Cook stem-side down in a steamer basket set over boiling water so that the tougher part cooks first. We find steaming gives a better texture than boiling.

To prepare sprouting broccoli, pull the larger bottom leaves off the thick main stalk. They are perfectly edible but, if the leaf stalks seem tough, fold the leaf in half lengthwise, then grasp the end of the stalk, pull it towards the tip and discard it. Cut off the flower heads, with about 8 cm/3 inches of stalk and the top tender leaves attached. Slice in half lengthwise if the stalk is thick.

With broccoli raab, you must be prepared for a lot of wastage: 450 g/1 lb makes only two or three servings. Discard the really tough stalks and large outer leaves, keeping the sprouting heads and young leaves.

Unfortunately, the purple colour of sprouting broccoli and Cape broccoli leaches out during cooking. In a stir-fry or pasta dish, for instance, this can muddy the appearance of the accompanying ingredients, so it is best to blanch either type briefly in boiling water before proceeding with the recipe.

## Cooking

Even if you tend to avoid creamy sauces and cooked cheese, it has to be acknowledged that dairy products go remarkably well with cauliflower. They seem to enhance its understated flavour, rather than mask it. Try as you might to venture down other culinary routes – Provençal or Indian, for example – no amount of tomatoes, garlic or curry spices seems to work quite as well. However, if you want to avoid the cliché of cauliflower cheese (though, well prepared, it can be good), try sautéing lightly steamed curds in a generous amount of unsalted butter with a clean-tasting assertive herb such as lovage (page 80). The addition of crisp-fried buttery breadcrumbs provides just the right amount of crunch and colour. Another way is to dip cauliflower florets in batter, such as the one used for Chiles Rellenos (page 20), deep-fry them until golden and serve with a freshly made tomato sauce enriched with a generous chunk of butter.

To show off its stunning good looks, 'Romanesco' cauliflower is best cooked simply and served without adornment. Unlike purple Cape broccoli, it keeps its vivid colour when cooked. The texture is surprisingly soft and creamy, despite the craggy curds, so keep the cooking time to a minimum – about 3–4 minutes – and preferably steam rather than boil.

Though good in a creamy gratin, really squeaky-fresh calabrese and sprouting broccoli, cooked until just tender and still a glowing emerald-green, are perfection on their own. Sometimes we anoint them with melted butter or good olive oil, sometimes with lemon zest, dried chilli flakes, mashed anchovies or freshly grated nutmeg. All taste very good. They are also delicious with large melting wafers of Parmesan strewn over them.

If your calabrese or broccoli is on the wane but still more or less edible, try chopping it into small florets and cooking with crisp-fried nuggets of bacon in a thick tortilla omelette; or add lightly steamed florets to a risotto, together with some finely chopped red chilli.

Also very good with pasta shapes, such as fusilli or conchiglie, is a mixture of small florets and thinly sliced stalks of calabrese briefly fried in 4–5 tablespoons of olive oil with diced pancetta, red onion, red pepper, garlic and lemon zest. Moisten with the pasta water and cook until the vegetables are tender, but still brightly coloured and crunchy. Stir in the pasta, remembering that there should probably be less pasta than vegetables, and scatter shavings of fresh Parmesan cheese over the top.

Equally delicious is broccoli and pasta – the ribbon type is best – moistened with a smoky green peppercorn and ricotta sauce. Gently fry about 450 g/1 lb of broccoli (stalks thinly sliced, florets cut vertically into two or three) with a smallish leek (quartered lengthwise and cut into 5 cm/2 inch shreds), half a red pepper, finely diced, and a generous amount of chopped flat-leaved parsley. When the vegetables are softened, stir in a 250 g/9 oz tub of ricotta mixed with an equal amount of whipping cream and 3–4 tablespoons of crushed green peppercorns. Stir the mixture until it is heated through, season it with salt and, if necessary, moisten with a little of the water in which you have cooked the pasta. Combine the vegetable mixture and the pasta in a warmed dish, toss them gently together and serve with plenty of Parmesan cheese.

Broccoli raab has a lovely chewy texture and rich flavour – slightly sweet, slightly bitter and almost meaty. We like it lightly cooked and served with goose or duck (page 80), especially with a spoonful of redcurrant jelly or chilli jam to balance the vegetable's slight bitterness. Broccoli raab is also good in a soupy stew with beans and sausage, or blanched, dried and braised with garlic and bacon.

## Nutrition

Calabrese and sprouting broccoli must top the list of disease-fighting vegetables. Over and over again, research projects reveal a strong and consistent pattern that a diet rich in vegetables, particularly green ones, decreases the risk of many cancers. Moreover, the evidence seems to indicate that cruciferous types, which include broccolis and cabbages, may be particularly effective in protecting against colon and thyroid cancers.

Both calabrese and sprouting broccoli are brimful with vitamin C and carotenes – antioxidants that offer protection against the harmful effects of free radicals, thus reducing the risk of cancer. An average serving contains well over the daily requirement of vitamin C and almost all the carotenes you need.

Sprouting broccoli and broccoli raab are particularly well-endowed with calcium, providing a quarter of the recommended daily dose, and broccoli raab is also a good source of potassium and magnesium. Sprouting broccoli contains an enormous amount of folate, one of the B vitamins which is essential for normal cell development.

Despite its unassuming appearance, cauliflower also provides valuable nutrients. Though comparatively low in carotenes and calcium, it contains reasonable levels of potassium, magnesium, folate and vitamin C.

## Varieties

We taste-tested all the varieties in late spring. They were steamed until just tender.

We found the outdoor-grown '**Late Purple Sprouting**' broccoli and '**White Star**', a late white sprouting broccoli, excellent, though for different reasons. The purple broccoli was tender, with a satisfyingly good flavour. The white type was dominated by leaf and stem, and had a denser texture and sweeter flavour.

'**Corvette**' (not shown), a calabrese which we grew in a polytunnel, has a sweet flavour but is not as complex as either of the sprouting types. Its texture was on the soft side, possibly a result of its pampered existence.

The overwintering cauliflower '**Snowbred**' had a well-rounded, satisfying sweetness, confirming our belief that this belongs to the royal family of winter vegetables.

We grew three types of broccoli raab in polytunnels. They were all bitter, but the strongest and least pleasant was '**Spring Raab**' (not shown). '**Sessantina Grossa**' (not shown), however, had a pleasant bitterness supported by both sweet and earthy flavours, and is a variety we will be trying again. '**Broccoletto**' fell somewhere between the two: bitter, but not unpleasantly so; sweet, but not quite up to the standard of 'Sessantina Grossa'.

## Pan-Fried Cauliflower with Lovage & Lemon▸

**Serves 4 as a side dish**

*1 small cauliflower, trimmed of leaves and central stalk*
*55 g/2 oz unsalted butter*
*5 tablespoons stale breadcrumbs*
*1 tablespoon olive oil*
*1 garlic clove, thinly sliced*
*3 tablespoons chopped lovage*
*½ teaspoon finely grated lemon zest*
*sea salt flakes*
*coarsely ground black pepper*

**Cauliflower benefits from the rich flavours of dairy products — in this case, butter — and contrasting crisp textures. Lovage is a neglected old-fashioned herb that is beginning to make an appearance once again in the shops. If you don't have any, use thyme or marjoram instead.**

**1** Steam the cauliflower over boiling water for 10–15 minutes, until it is reasonably tender but not disintegrating. Divide into florets and set aside.

**2** Melt half the butter in a frying pan over moderate to high heat. When it is foaming, add the breadcrumbs and fry until crisp. Transfer to a small bowl and wipe out the pan.

**3** Heat the oil and the remaining butter over moderate heat. Add the garlic, lovage and cauliflower. Fry for 3–4 minutes, tossing gently, until the cauliflower is flecked with gold.

**4** Transfer the contents of the pan to a heated serving dish. Add the breadcrumbs, lemon zest, a good pinch of sea salt flakes and several grindings of black pepper. Pour over any juices remaining in the pan and toss carefully to mix.

## Grilled Duck Breast with Broccoli Raab & Chilli Jam

**Serves 4**

*3 Barbary duck breast fillets, about 280 g/10 oz each*
*sea salt flakes*
*freshly ground black pepper*
*900 g/2 lb broccoli raab, tough stalks and outer leaves removed*
*knob of butter*
*chilli jam, to serve*

**This is one of those accidental recipes — we happened to have duck breasts in the freezer, a pot of chilli jam and some broccoli raab. It is a brilliant combination of sweet, hot and bitter. We served it with baked breadfruit, which, rather remarkably, we also happened to have. It was a delicious foil to the other flavours. Plantains, celeriac and potato mash, or steamed new potatoes would be good alternatives.**

**Chilli jam is sold in some of the smarter food shops. You could substitute redcurrant jelly mixed with finely chopped roasted chilli.**

**1** Rub the duck skin with salt and season the breasts on both sides with salt and pepper. Place skin side down in a grill pan and cook under a preheated hot grill, about 15 cm/6 inches from the heat source, for 10 minutes. Turn the breasts over and grill for 10–15 minutes more, until the skin is crispy and the flesh still juicy and slightly pink. Leave to rest in a warm place.

**2** Have ready a large pan of boiling salted water. Plunge in the broccoli raab and cook for about 5 minutes, until it is just tender. Drain thoroughly, squeezing out all the moisture, then return to the pan, season to taste, and add the butter.

**3** Thinly slice the duck breasts at an angle. Divide the broccoli raab between four warmed plates and arrange the duck slices on top, pouring over any juices that have accumulated. Serve with a spoonful of chilli jam and a starchy vegetable of your choice.

# LEAVES & STEMS

ORIENTAL GREENS LEAFY GREENS

HERBS BITTER LEAVES

CABBAGES & KALES

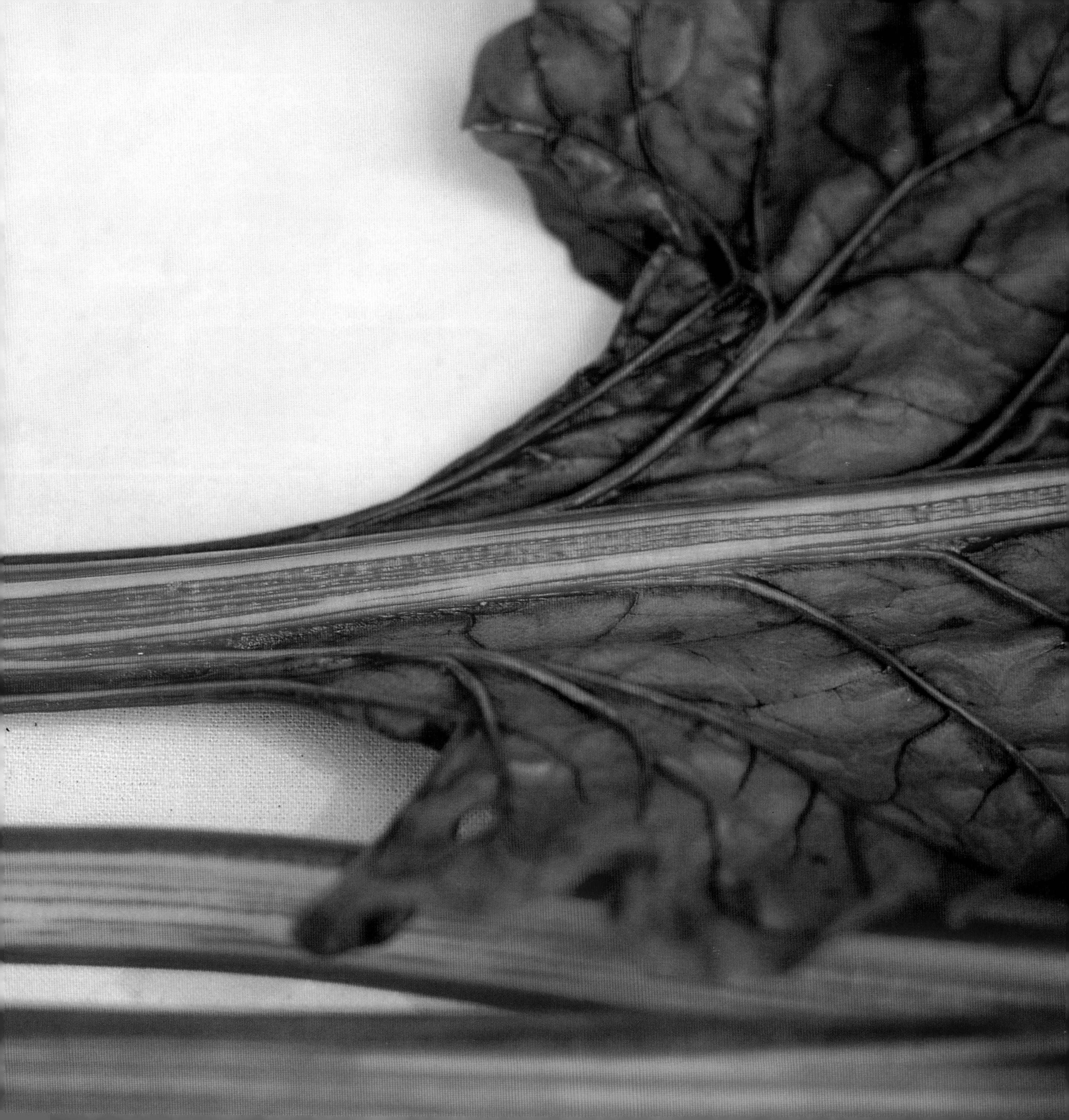

# ORIENTAL GREENS

As a group, oriental greens are notable for their generosity. They provide an impressive display of colours – the wine-red of some mustard leaves, the almost unnaturally dark green of tatsoi, and the delicate pale ivory of the self-blanching Chinese cabbages. They also contribute an abundant selection of textures, from the bold crunchiness of the leafy bok choys to the chewiness of stemmed varieties such as choy sum and Chinese broccoli. To be on the receiving end of such munificence is a satisfying experience, leaving cooks and their beneficiaries with a sense of well-being.

## An identity crisis

Oriental greens do, however, suffer from an identity crisis. Scientific nomenclature is in a permanent state of flux, and assigning the greens correct botanical names seems to be a hotly contested issue among the experts. Fortunately, cooks do not have to participate in these esoteric discussions, though they have an almost equally onerous task in figuring out the colloquial names.

One of the difficulties is that English names are often transliterated from one or more Asian languages, resulting in sometimes subtle, though oftentimes significant, differences in spelling and pronunciation. For instance, 'bok choy' and 'pak choi' from the Cantonese and 'shagkushina' from the Japanese all refer to the same vegetable. In other instances different English names are attached to a vegetable without reference to its Asian names. Chinese cabbage, for example, is called not only Chinese leaves but also celery cabbage, michihili and Napa cabbage.

## Basic botany

Although there are exceptions, water spinach and Malabar spinach for example, the majority of oriental greens are brassicas. As such, they have the characteristically pungent and bitter flavours that are caused by the sulphur-containing chemicals known as glucosinolates.

The leaves of many of these plants consist of two parts: a thin green flattened blade, and a petiole or stalk to which the blade is attached. The blade itself has a central mid-rib, which is simply an extension of the stalk. Branching out from this mid-rib are veins that give the blade structure and support. The leaves themselves are arranged in a whorl around the plant's compressed central stem, growing close to ground level. Older leaves are on the outside of the whorl, while new ones emerge from the inside.

When a plant starts to flower (or 'bolt', in vegetable gardening terms), the central stem extends upwards, causing the leaves to become spaced more or less evenly along its length. Although oriental greens are mostly grown for their leaves, some varieties are cultivated for their flowering stems and tend to flower quickly after planting. Even when leafy varieties bolt, the young stems will still be tender and pleasant to eat.

## Cultivation and harvesting

In general, the leafy brassicas prefer cool growing conditions, and do best in the shortening days of mid-summer and early autumn. Spring-sown plants, especially if exposed to excessive cold at the seedling stage, may bolt. To overcome this, use bolt-resistant varieties or delay sowing until mid-summer. To extend the harvesting season, the more cold-tolerant varieties can be cultivated through mild winters, especially when given the protection of a plastic tunnel or a greenhouse.

Home-growers have a distinct advantage over shoppers in that leafy greens can be harvested at different stages of maturity, producing, in effect, different vegetables from the same plant. Cut when they are young seedlings, the plants yield a tender crop of diminutive leaves perfect for salads and sandwiches. They then grow back in full force, and are quickly ready for a repeat harvest. Or cutting can be delayed until the plant is more mature. At this stage all the leaves can be cut about 4 cm/1½ inches above ground level and then left to regrow. Alternatively, the mature outside leaves can be individually harvested, leaving the younger interior leaves to carry on growing.

Whole plants, too, can be harvested all in one go after they reach full maturity, a stage that brings out the full potential of the heading varieties, like Chinese cabbage.

Oriental greens grown for their stems and flowers also prefer a cool climate. Hardier varieties can be over-wintered outdoors in milder areas, or in a tunnel or greenhouse where it is colder.

To get the best results, the stem should be cut just before or just after flowering begins. Three or four leaves should be left at the base of the stem, and new, though thinner, stems will develop.

## Buying and storing

Though the home-grower has access to a greater selection of oriental greens, the shopper still has an exciting choice, especially from stores catering to the Asian communities. The larger supermarkets also carry a reasonable selection, particularly during the Chinese New Year.

When buying Chinese cabbage, look for firm heads that feel heavy for their size. Don't worry about tiny black flecks on the leaves – these are normal and harmless – but do reject heads with tattered outer leaves or those that look wilted. Stored in a plastic bag in the salad drawer of the fridge, a cabbage in good condition will keep for two to three weeks. Do not wash Chinese cabbage before storing as this will speed up deterioration.

Other greens, such as bok choy, tatsoi (flat cabbage) and choy sum (flowering cabbage), are less accommodating. Most of them need to be used within a day or two of purchase, although Chinese broccoli can be kept for up to a week. Buy only crisp, firm specimens with undamaged leaves. Store them unwashed, wrapped in damp paper towel and then in a plastic bag in the salad drawer.

## Preparation

To prepare Chinese cabbage, simply cut off the base, rinse the leaves in cold water and dry them thoroughly. If you need only a portion of cabbage, cut out a lengthwise segment, then slice the base off this part only. The leaves and stalk can either be cut across or sliced lengthwise into strips.

In the case of bok choy and other varieties with densely packed fleshy-based leaves, separate the leaves and rinse them in several changes of water to remove any grit. For stir-frying, it's best to cut the green leafy part from the fleshy base, as the base needs longer to cook. Tender baby bok choy, however, can be cooked whole with the leaves intact. Rinse the whole plant under running water, then slice lengthwise into three or four segments.

To prepare Chinese broccoli, separate the leaves from the stems and rinse briefly. If the stems are very tough, peel them before use; they may need blanching, too, if they're intended for a stir-fry.

Choy sum can be left whole – there is no need to peel the stems. Soak it in plenty of cold water, making sure you loosen any grit at the base of the stems.

Mizuna and mibuna need very little preparation. If necessary, rinse the leaves briefly in cold water and dry them thoroughly before using in a salad or stir-fry. There is no need to cut off the stalks – they are tender and juicy.

## Cooking

Oriental greens are vivid and emphatic – peppery, sometimes slightly bitter, with a clean, refreshing taste. They combine well with fish, seafood and poultry, and yet are assertive enough to stand alone.

Elaborate preparation is unnecessary, and steaming or stir-frying are sufficient to bring out their personalities. Chinese cabbage can be boiled in the same way as ordinary cabbage, but prolonged contact with water tends to dilute the delicate flavour. Best results are obtained from the intense direct heat of the wok. Try it in a stir-fry with blackened pork (page 90).

Used raw, oriental greens make a welcome addition to salads, especially in the winter when their pungency and colour provide a contrast to the insipidity of most greenhouse-grown lettuces.

These plants require a sympathetic touch when it comes to dressings and seasonings – soy sauce, oyster sauce, toasted sesame oil, garlic and ginger are more in keeping with their oriental nature than the more robust Mediterranean oils and condiments.

## Nutrition

Oriental greens are packed with vitamins and minerals, containing over three times more calcium, potassium and iron than ordinary cabbage, and over six times more of the antioxidant beta-carotene, widely believed to protect against heart disease and cancer. They provide large amounts of vitamin C and E, which are also antioxidants implicated in protection against certain cancers. Even more in their favour, these greens contain 'new' antioxidant carotenes such as lutein and zeaxanthin, which not only boost the immune sysem but are also thought to be directly related to sharpness of vision. The darker green varieties, such as mustard greens, tatsoi and bok choy, contain significant amounts of folate – vital for growth and development.

## Varieties

Levels of pungency and bitterness, a common characteristic of these greens, vary depending on variety, maturity and the growing environment, so it really is a matter of 'grow and see'. We conducted tasting trials on plants grown in tunnels throughout the winter. We tasted the outer leaves of the leafy varieties and the stems, leaves and flower buds of the stemmy types.

### Chinese cabbage

There are two basic types of Chinese cabbage (*Brassica rapa* Pekinensis Group): firm-headed and loose-headed. Firm-headed cabbages are further divided into short, squat barrel-shaped varieties, sometimes called Napa cabbage, and taller cylindrical

'Green Lance'
'Nabana'
'Hon Tsai Tai'
'Autumn Poem'

ones known as michihili. Loose-headed types range from the really open and floppy to those that have loose fluffy tops like a Cos lettuce. Mature heads, especially those that are tightly packed, are self-blanching, producing succulent, ivory-coloured inner leaves with smooth, broad mid-ribs.

'**Kasumi**' (not shown), a barrel-shaped variety, is delicious cooked, and has a nice full flavour when raw. Coarsely chopped, the raw leaves make a crisp, cool bed on which to serve a whole fish, or barbecued duck or chicken, and they also make a welcome alternative to lettuce in winter salads. '**Jade Pagoda**' (not shown), a michihili type, is mild to the point of being bland when raw, but takes on a pleasant gentle flavour when cooked. In contrast, '**Ruffles**', a fluffy-top type, is bitter both raw and cooked.

### Bok choy, pak choi

The leaves of bok choy (*B. rapa* Chinensis Group) have white or pale green, fleshy, spoon-shaped stalks. A relatively tall-growing, white-stalked variety with a great flavour is '**Joi Choi**' (not shown). It has a crisp juicy stalk and mid-rib, and is delicious either raw or cooked. It makes a brilliant salad with crisp cucumber and beansprouts (opposite). It is also good in soups. An attractive, compact variety is '**Mei Qing Choi**'. The stalk and mid-ribs are pale green, and are not as crisp or flavoursome as that of 'Joi Choi', whether raw or cooked.

### Tatsoi, tatsai, tah tsai

The leaves of tatsoi (*B. rapa* Narinosa Group) have oval, very dark green blades attached to a somewhat fleshy stalk. The plant is also known as rosette pak choi, because of its flattened, ground-hugging growth habit. When cooked, it has a crunchy texture and a pungent, pleasantly strong flavour. Try it stir-fried with garlic, ginger and soy sauce, and serve it with seared salmon or tuna. Raw, it has no pungency and tastes almost grassy, but it is useful for adding texture and colour to salads.

### Mibuna

Mibuna (*B. rapa* Japonica Group) has an upright growth habit, with slender dark green leaves, white stalks and midribs. When cooked, '**Green Spray**' is both bitter and pungent. The texture, although somewhat stringy, is pleasantly chewy. Eaten raw, it is slightly pungent, with some grassy flavour in the background.

### Mizuna

Mizuna (*B. rapa* Japonica Group) has an upright growth habit, with bright green deeply cut blades and thin white stalks and mid-ribs. This is outstanding for eating raw, with just a hint of pungency and a gratifying chewy, crunchy texture. With their feathery shape and bright colour, the leaves really add sparkle to a salad. They are also good cooked. The mature leaves are delicious quickly tossed with pasta or stir-fried prawns or chicken.

### Mustard greens

The leaves of mustard greens (*B. juncea*) have a slightly rough surface, which can be a distraction when they are eaten raw. Some varieties, like '**Red Giant**', have red-tinted leaves, while others are completely green. Depending on the plant's variety and age, and the time of year, pungency ranges from the fiery heat of '**Green in the Snow**' (not shown) to the glowing warmth of 'Red Giant'.

### Flowering and stemmed greens

One of the best known of the greens grown for their stems is Chinese broccoli, also known as kailan, gai lan and various other transliterations. '**Green Lance**' (*B. oleracea* Alboglabra Group) has an agreeable mustard flavour and is somewhat bitter, though not unpleasantly so.

Another bitter variety is '**Nabana**' (*B. rapa* Pekinensis Group), a flowering Chinese cabbage). It has light green crinkled leaves and thick stems, and is a very robust grower. '**Autumn Poem**' (*B. rapa* Parachinensis Group), a flowering choy sum, was a trifle disappointing. We found its bitterness so overpowering that it masked any positive characteristics.

Our favourite variety is '**Hon Tsai Tai**' (*B. rapa* var. *purpurea*), also known as purple-flowered choy sum. The plant is purple-tinged, and the stems are thinner than those of 'Green Lance' and 'Nabana'. The flavour is exceptional, with a slight but pleasant bitterness contrasted against a hint of asparagus. Allow some of the bright yellow flowers to bloom before harvesting, and you have the perfect vegetable – delicious and beautiful.

# Salad of Oriental Greens, Cucumber & Beansprouts

**Serves 4**

*40 g / 1½ oz macadamia nuts or large peanuts*
*½ teaspoon muscovado sugar*
*½ teaspoon salt*
*finely grated zest of 1 lime*
*10 cm/4 inch piece of cucumber*
*175 g/6 oz mixed baby oriental leafy greens, such as baby bok choy, tatsoi, mizuna and mustard leaves, torn into bite-sized pieces*
*85 g/3 oz mung beansprouts*

**For the dressing:**

*1 tablespoon rice wine vinegar*
*1 teaspoon shoyu or tamari*
*salt and freshly ground black pepper*
*4 tablespoons macadamia oil or extra-virgin sunflower oil*
*1 shallot, finely chopped*

**Serve this clean, fresh-tasting salad either as a starter or as an accompaniment to grilled fish or poultry. The combination of chopped nuts, lime zest, salt and sugar adds sweet-sharp crunch. The dressing is deliberately kept light; if you don't have macadamia or extra-virgin sunflower oil, use a light olive oil instead.**

1 First make the dressing: whisk together the vinegar, shoyu or tamari, and salt and pepper to taste. Slowly pour in the oil, whisking constantly until smooth. Stir in the shallot and set aside for 30 minutes. Check the seasoning and whisk again before using.

2 Meanwhile, preheat the oven to 180°C/350°F/gas 4. Put the nuts in a small roasting tin and roast for 10–12 minutes, until golden. If they have skins, remove them by rubbing with a clean, dry tea towel. Chop the nuts quite finely and mix with the sugar, salt and lime zest. Set aside.

3 Peel the cucumber and halve lengthwise. Scoop out the seeds and thinly slice the flesh diagonally. Place in a bowl with the greens and beansprouts.

4 Add the dressing to the leaves and toss gently to coat. Divide the salad between individual plates and sprinkle the nut mixture over the top. Serve immediately.

# Blackened Pork on Chinese Cabbage

**Serves 4**

*1 kg/2¼ lb boneless belly pork in one piece, with the rind left on*
*4 tablespoons groundnut oil*
*150 ml/¼ pint chicken stock*
*5 cm/2 inch piece of fresh ginger root*
*1 tablespoon shoyu or tamari*
*1 tablespoon dry sherry or rice wine*
*1 teaspoon nam pla fish sauce*
*800 g/1¾ lb Chinese cabbage, leaves and stalks separated and cut into bite-sized pieces*
*4 good-sized spring onions, green and white parts separated, sliced at an angle into 2 cm/¾ inch pieces*
*115 g/4 oz mangetouts, halved lengthwise at an angle*
*½ teaspoon salt*

**For the marinade:**

*3 tablespoons sugar*
*3 tablespoons dry sherry*
*1 tablespoon shoyu or tamari*
*4 tablespoons Hoisin sauce*
*5 cm/2 inch piece of fresh ginger root, squeezed in a garlic press*
*1 teaspoon salt*
*½ teaspoon ground cinnamon*
*15 black peppercorns, crushed*
*3 cloves, crushed*
*1 star anise pod, crushed*

**The ingredients list here may seem daunting, but the dish is very easy. The pork can be marinated, roasted and sliced the day before. The final stage of cooking takes no more than 10 minutes.**

**1** Using the tip of a very sharp knife, score the pork rind diagonally at 1 cm/½ inch intervals. Cut the meat into four strips measuring approximately 15 x 5 cm/6 x 2 inches.

**2** Combine the marinade ingredients and pour the mixture into a shallow dish in which the meat can sit snugly in a single layer. Add the pork strips, rubbing the marinade into the crevices. Cover and leave at room temperature for 1 hour, turning now and again.

**3** Preheat the oven to 220°C/425°F/gas 7. Line a roasting tin with foil and sit a rack in it. Place the pork, rind side down, on the rack and put the tin in the top part of the oven. Reserve the marinade for basting. Roast for 15 minutes, then reduce the oven setting to 180°C/350°F/gas 4. Roast for 30 minutes, basting frequently with the marinade, then turn the pork over and roast for another 30 minutes, continuing to baste. Allow to cool completely, then slice at an angle into 2.5 cm/1 inch pieces.

**4** When you are ready to complete the dish, heat 1 tablespoon of the oil in a large frying pan over high heat. Add the pork slices and fry until crisp and blackened. Pour in the stock, scraping up any nice sticky sediment. Cook for another minute, then set aside and keep warm.

**5** Squeeze the ginger in a garlic press. Combine the ginger juice with the shoyu or tamari, sherry, 1 tablespoon of water and the fish sauce. Preheat a wok or large frying pan over high heat. Pour in the remaining oil and, when it is almost smoking, add the cabbage stalks and the white parts of the spring onions. Stir-fry for 2 minutes, then add half the ginger mixture and fry for another few seconds. Throw in the cabbage leaves, the mangetouts, the green spring onion and the remaining ginger mixture. Stir-fry for 1–2 minutes – the cabbage should still be crisp and retain its colour. Season with salt.

**6** Divide between individual serving plates and arrange the meat and any juices on top.

# LEAFY GREENS

Leafy greens tend to be regarded in much the way that teenagers view their parents: dependable, boring and to be taken for granted, perhaps even bitter, gritty and tough. It is perhaps worth noting, then, that in societies where the older generation is treated with respect and affection, leafy greens, too, are deservedly appreciated. In places as far-flung as the Mediterranean, the Middle East, the Caribbean and the Orient, they are consumed with relish and enjoyment.

Leafy greens are, in fact, exciting and adventurous. Their multicultural backgrounds make them tickets to exotic destinations and they show enough versatility, as far as the cook is concerned, to rival any star of the vegetable world.

Take as an example their colours. The extroverted burgundy-red stems of '**Rhubarb Chard**', and the dazzling yellow and orange stems of newer chard varieties, lend a carnival atmosphere to any vegetable plot. Flavours are intriguing, too, ranging from the satisfying earthiness of the amaranth (or callaloo) to the refreshing mild astringency of New Zealand spinach. Texture adds more drama and surprise. Crisp, crunchy Swiss chard stems are a vegetable in their own right, contrasting well with the velvety leaf; while the thick, somewhat abrasive, leaves of New Zealand spinach line up in an orderly fashion along a succulent stem. Malabar spinach is not nicknamed 'the slippery vegetable' for nothing – it really does have a melt-in-the-mouth quality.

Leafy greens have vastly different growing requirements (see Varieties, page 95). For example, some are tropical and where the weather is cooler need to be grown in a polytunnel or greenhouse, while others do well outdoors even in temperate climates.

## Buying and storing

If you don't grow your own leafy greens, it's worth seeking out the more exotic varieties in Asian or Middle Eastern food shops. For some reason, Swiss chard is not always easy to find, but at last some supermarkets are beginning to stock different varieties. Farmers' markets are also good hunting grounds.

When buying leafy greens, look for crispness and bounce. Fresh greens should positively squeak with life.

Malabar spinach
New Zealand spinach
Rainbow chard
Amaranth 'Red Leaf'
Water spinach

A moist cool atmosphere will temporarily postpone wilting, so once you've got your greens home, store them unwashed and untrimmed, wrapped in damp paper towels in a roomy plastic bag in the salad drawer of the fridge. Use them within a day or two, and wash and trim them just before you're ready to cook.

## Preparation

Most leafy greens need dunking in several changes of cold water to remove grit. Don't discard the stems as, in most cases, they are tender and juicy, adding texture to the finished dish. The stems will need cooking for a few minutes longer and should be stripped from the leaves so they can be first in the cooking pot. To remove Swiss chard stems, fold the leaf in half so that the upper surfaces touch, then grasp the stem and pull it firmly towards the tip. The smaller leaves of most varieties can be left whole, or torn into bite-sized pieces. Stack larger leaves and cut them across into ribbons, using a very sharp knife to avoid bruising.

Swiss chard, amaranth and New Zealand spinach can be blanched before cooking. Blanching is useful before cooking a stir-fry, for instance, as it removes some of the moisture and will also intensify the colour. Plunge the greens into a large pan of boiling water for a minute or two, then briefly rinse them under cold running water to stop further cooking. Blanched greens also make richly flavoured fillings for pies, tarts and parcels. Try them mixed with ricotta cheese and Parmesan in home-made ravioli, or with mushrooms in a filo tart. They can also be chopped and added to thick chunky omelettes or soufflés.

## Cooking

Although tiny tender leaves may be used raw, most leafy greens need cooking. Keep this brief, though, or the delicate texture and flavour will be lost, as will the vitamin content. Being composed mostly of water, leafy greens dramatically reduce in bulk once cooked. A huge bunch of uncooked leaves may look like far too much, but it will cook down to about half its original volume.

Leafy greens are best cooked simply, either by boiling or steaming, or stir-frying. There's no need to add any extra water when boiling – there will be enough from washing and the greens will also eventually give up their own moisture. Leaf beets and New Zealand spinach, like ordinary spinach, have an affinity with dairy products, so if you are not unduly worried by fatty indulgences, stir a good-sized pat of unsalted butter or a dollop of cream into the cooked and drained greens.

For stir-frying greens to serve with pasta, for instance, heat a generous amount of olive oil in a large pan over moderate to high heat. Add your chosen greens, thick stalks first if you are using them, with some chopped garlic. Cover and cook the stalks for a minute or two, then add the leaves. When they've wilted and reduced a little, uncover and toss them until they are just tender. Add seasonings, including some finely grated lemon zest and 2–3 tablespoons of chopped fresh dill if you like. Toss with cooked pasta shapes. This is even more delicious sprinkled with crisp-fried breadcrumbs. Serve with plenty of freshly grated Parmesan cheese.

For an oriental stir-fry, try the recipe for amaranth and enoki mushrooms on page 96, or simply sizzle slivers of garlic, ginger and perhaps some chopped chilli in groundnut oil, throw in the greens and cook as above. Season with soy sauce and a splash of toasted sesame oil. For an Indian version, add ground turmeric and cumin, and leave out the soy sauce and sesame.

Leafy greens also make good soup. Tropical greens, such as water spinach or Malabar spinach, are beautiful in an oriental-style soup (page 98), the delicate leaves floating like lily pads in a well-flavoured clear broth. In contrast, the more robust amaranth leaves hold their own in the classic Jamaican pepperpot soup – a gutsy brew of amaranth cooked with okra, yams, chillies, beef, pig's tail and coconut milk.

## Nutrition

Like all dark green vegetables, leafy greens are packed with vitamins and minerals, particularly antioxidants – a group containing carotenes, vitamin C and vitamin E, all of which are strongly implicated in protection against certain cancers. Malabar spinach and amaranth, for example, have astonishingly high levels of both vitamin C and beta-carotene.

All leafy greens are an excellent source of folate, one of the B vitamins. Contrary to popular belief, they are not a particularly rich source of iron, although they do contain useful amounts, together with calcium, magnesium and small amounts of zinc.

The downside of leafy greens is their oxalic acid content – it is this that makes your mouth pucker. Oxalic acid combines with calcium and iron, forming insoluble compounds that pass unabsorbed through the gut and may also accumulate as kidney stones. In the quantities normally eaten, however, oxalic acid is rarely a matter for concern and any worries should be offset by the benefits of the high concentrations of antioxidants present.

## Varieties

Seed is readily available and it is well worth experimenting with the more unusual varieties, even tropical ones. Although several go by the name of 'spinach', in fact they belong to completely different botanical families.

### Water spinach

Despite the name, some types of water spinach *(Ipomoea aquatica)* grow perfectly well in soil. It is a common ingredient in both Indian and Chinese dishes. Needing warmth and protection in cooler climates, it grows as a short vine with hollow stems. Both leaves and stems can be eaten, and they turn a lively green when cooked. The texture is a touch mucilaginous and the leaves leave the mouth slightly dry. Teaming with pungent ingredients, however, counteracts the dry-mouth effect. Try it stir-fried with garlic, ginger and shrimp paste or fermented tofu.

### Malabar spinach

Also known as Ceylon spinach and 'slippery vegetable', Malabar spinach *(Basella rubra* and *B. alba)* is another tropical variety that can be grown under cover in a cool climate. The plant produces impressive vines of succulent, heart-shaped, shiny green leaves with thick stems. It has a mild fresh green flavour that is good in soup. Both leaves and stems can be used. Take care not to overcook them, or the interesting slippery quality may become slimy.

### New Zealand spinach

New Zealand spinach *(Tetragonia tetragonioides)* has sprightly, slightly abrasive, triangular leaves with crisp round stalks. It is a favourite with gardeners, since it can withstand heat and drought, and is a winner with cooks for its full flavour. Once harvested, it keeps reasonably well and has less tendency to bruise. Unlike ordinary spinach (*Spinacia deracea)*, it does not reduce in bulk much when cooked.

### Leaf beets

Leaf beets *(Beta vulgaris* subsp. *cicla)* are closely related to beetroot. These leafy greens offer two vegetables for the price of one, since the stems can usually be cooked separately and eaten as a vegetable in their own right.

**Swiss chard** has a broad white stem with a dark green crinkled leaf. In contrast, **perpetual spinach** has a narrower light green stem and a smooth light green blade, while '**Rhubarb Chard**' (also known as ruby chard) has red stems and red-tinged blades. **Rainbow chard** produces plants with stems that come in flamboyant shades of red, pink, orange and yellow. '**Bright Lights**' (not shown) is another chard with broad stems that come in a range of bright and pastel colours. One of the best dishes for making the most of the magnificent colour is a salad of crisp-cooked multi-coloured stems and soft chunks of avocado on a bed of dark green lettuce (page 96). Alternatively, try braising the stems and chopped leaves with garlic and chilli flakes, and topping them with lightly cooked young broad beans.

### Amaranth

Amaranth *(Amaranthus tricolor)* is thought to be a native of both the Andes and India. This dual nationality seems to have conferred upon it a certain degree of tolerance, and it has easily become assimilated into the varied cooking styles of India, China and the West Indies (where it is known as callaloo). Although it is associated with tropical climates, amaranth grows quite well farther north in the warmer summer months. One particularly attractive variety is '**Red Leaf**', which has oval to heart-shaped leaves. They have a light green fringe surrounding a splashy deep red centre, which bleeds along the veins.

Amaranth is one of the most satisfying greens to cook. It has a lovely earthy flavour with no acidic aftertaste, despite its reputedly high oxalic acid levels. It stands up well to cooking with spices, and is good in risottos, stews and stir-fries.

## Rainbow Chard & Avocado Salad ▸

**Serves 4**

*450 g/1 lb rainbow chard stems without leaves*
*1 large avocado*
*4 good handfuls of robust salad leaves, such as baby spinach, 'Little Gem' or lamb's lettuce*
*12 oil-cured black olives*
*sprigs of flat-leaved parsley, to garnish*

**For the lemon dressing:**

*1 garlic clove*
*generous pinch of sea salt flakes*
*freshly ground black pepper*
*2 tablespoons lemon juice*
*6 tablespoons extra-virgin olive oil*

**For a spectacular psychedelic effect, choose a mixture of coloured stems — pillar-box red, zinging cyclamen and brilliant chrome yellow. Soft satiny avocado contrasts well with the crunchy stalks. This salad is particularly good served while the stems are still slightly warm.**

**1** Using a very sharp knife, trim a sliver from each side of the chard stems – the edges can sometimes be a little stringy. Then diagonally slice the trimmed stems across at an angle into neat 2 cm/¾ inch pieces.

**2** Make the dressing: bash the garlic and sea salt with a pestle and mortar. Mix with the pepper and lemon juice, then whisk in the oil until thick and smooth.

**3** Plunge the prepared chard stems into a large pan of boiling salted water. Bring back to the boil and blanch the stems for a couple of minutes – no more or the bright colours will fade. Drain immediately and toss with the dressing.

**4** Just before serving, finely dice the avocado flesh.

**5** To serve, arrange the salad greens on individual plates. Scatter the stems over the top and add the avocado and olives. Spoon over the dressing and garnish with a sprig of flat-leaved parsley.

## Stir-Fried Amaranth & Enoki Mushrooms with Jasmine Rice

**Serves 2**

*225 g/8 oz amaranth leaves*
*85 g/3 oz enoki mushrooms*
*115 g/4 oz jasmine (Thai fragrant) rice*
*2 tablespoons groundnut oil or extra-virgin sunflower oil*
*1 garlic clove, very finely chopped*
*15 mm/¾ inch piece of fresh ginger root, very finely chopped*
*½–1 fresh red chilli, deseeded and very finely chopped*
*½ small red pepper, deseeded and finely diced*
*2–3 tablespoons chicken or vegetable stock, or water*
*2 teaspoons tamari or shoyu*

**Dark green amaranth leaves, splashed with beetroot red, make a technicolour stir-fry. The rice will become tinged with a beautiful pink. If the amaranth is late-season, it may need first to be blanched briefly in boiling water — it depends if you like your leafy greens chewy or soft.**

**Enoki mushrooms are a fascinating fungus — they grow in clumps and have tiny round heads with long thin stalks. Larger supermarkets stock them, but if you can't find any use thinly sliced oyster or shiitake mushrooms instead.**

**1** Remove any tough stalks from the amaranth and chop the leaves coarsely. Cut the root clump from the mushrooms. Cook the rice according to the packet instructions.

**2** Meanwhile, heat the oil in a large frying pan and gently fry the garlic, ginger, chilli and red pepper for 2 minutes, stirring.

**3** Add the amaranth and mushrooms, pushing the leaves around until they start to wilt. Cover and cook over low heat for 5 minutes, stirring occasionally. Add a little stock or water if the vegetables start to get dry.

**4** Uncover, increase the heat and stir in the tamari or shoyu. Stir-fry for 2–3 minutes, until the leaves are cooked to your liking. Divide the cooked rice between two plates and serve with the vegetables on top.

# Chinese Spinach & Water Chestnut Soup

**Serves 4**

*225 g/8 oz Malabar spinach, leaves and stems separated and chopped into bite-sized pieces*

*half a 225 g/8 oz can water chestnuts, drained and thinly sliced*

*1 teaspoon light soy sauce*

*3 spring onions, green parts included, sliced at an angle*

*salt and freshly ground black pepper*

*2 tablespoons chopped garlic chives*

*a few garlic chive flowers, to garnish*

**For the Chinese stock:**

*450 g/1 lb chicken wings, coarsely chopped*

*450 g/1 lb pork spare ribs, coarsely chopped*

*2.5 cm/1 inch piece of fresh ginger root, thickly sliced*

*2 fat spring onions, halved lengthwise*

*1–2 tablespoons rice wine or dry sherry*

**Water chestnuts provide a crisp contrast to the silky spinach. Use ordinary spinach if you can't get hold of Malabar spinach, but discard the tough stems. Garlic chives produce minute white flowers which look beautiful scattered over dark green leaves, but don't worry if you don't have any — the soup is still delicious without.**

**You need a clear flavourful stock for this soup. If you don't want to make Chinese stock, use 1.2 litres/2 pints of vegetable bouillon or low-sodium chicken stock made with powder or a cube. Simmer it for 15 minutes with a couple of slices of fresh ginger root. Drain and discard the ginger before using.**

**1** First make the Chinese stock: put the chicken and pork in a saucepan with 1.3 litres/ 2¼ pints of water. Slowly bring to the boil, skimming off any scum. Add the ginger, the spring onions and the rice wine, if you are using it. Reduce the heat to the gentlest of simmers and cook, uncovered, for 1–2 hours. Strain through a colander, then again through a muslin-lined sieve. You will need 1 litre/1¾ pints.

**2** Bring the stock back to a gentle boil, add the spinach stems and cook for 2 minutes. Then add the leaves, the water chestnuts, the soy sauce, the spring onions, and salt and pepper to taste. Simmer for another 2 minutes and stir in the garlic chives. Ladle into bowls and sprinkle with a few garlic chive flowers.

# HERBS

Culinary herbs are prized for the flavours they bring to a dish. The aromatic *fines herbes*, such as rosemary and thyme, are so robust that they are used only in small amounts. Some of the leafier herbs, however, are less potent, and may be used in much larger quantities. Orache, nasturtiums and purslane, for example, are almost vegetables in their own right, contributing body, texture and colour as well as flavour.

Herbs offer a true melting pot of scents and flavours, emanating from essential oils stored in the leaves and stems. The oils are released when the herb is heated or crushed, but since the oils are volatile, their flavour is often fleeting or can be destroyed by overcooking. Essential oils are not the whole story, however. For example, oxalic acid is responsible for the sourness of sorrel, while the glucosinolates in rocket result in a pungent hot flavour.

## Cultivation and harvesting

It is difficult to generalize about herbs, since each type requires a slightly different growing environment. Most do best in a sunny spot, though buckler-leaved sorrel is fine in the shade. In contrast, basils need plenty of both heat and light, and may be much better off in a greenhouse or polytunnel. They are also frost-sensitive and should not be planted out until danger is past. Purslane is another frost-sensitive herb. In comparison, orache, rocket and parsley are more cold-tolerant and grow happily outdoors.

Herbs are easy to grow and do perfectly well in any well-drained soil. If you don't have a garden, you can try them in containers on a patio or balcony, or a south-facing windowsill in the house. Be careful when growing them this way: while some, such as nasturtiums, are ideal for pots and hanging baskets, tall-growing varieties like orache and epazote are less suitable and need their growing tips pinched out to encourage bushiness.

## Buying and storing

Supermarkets stock a large range of herbs, mostly sold in small plastic packets or pots. More generous bunches can be found in street markets and stores catering to the Middle Eastern, Latin American and Asian communities.

Leafy herbs should look as if they have just been harvested – dewy-fresh and exuberant. Reject any that are jaded or limp, or that have yellowing leaves or brown patches. If possible, use herbs on the day of purchase, otherwise loosely wrap them in damp paper towels, unwashed, and store them in a sealed plastic box in the salad drawer of the fridge. Depending on freshness, they will keep for 2–3 days. Large bunches of parsley will last for several days if placed in a jug of water with a plastic bag loosely covering the leaves.

## Preparation

The leaves of most varieties will need to be stripped from their thicker stems. Depending on the herb and the recipe, the stems need not be wasted, for they are a source of concentrated flavour and nutrients. Finely chopped parsley stems, for example, can go into a puréed soup or sauce; or whole stems can be tied in a bundle and fished out of the dish before serving.

The leaves are chopped, pounded or left whole, depending on the recipe. More than any other leafy plants, herbs need to be chopped with a very sharp blade, otherwise you will simply bruise them rather than cutting them cleanly.

## Cooking

Because prolonged exposure to heat tends to destroy their flavour, leafy herbs are best used raw or added to cooked dishes just before serving. A good handful of raw basil, for example, puréed with olive oil or light stock, makes a concentrated dressing or sauce that is perfect with fish or a warm vegetable salad. Alternatively, add capers and lemon juice to make *salsa verde*, a piquant green sauce, or Parmesan and pine nuts for pesto sauce.

Mediterranean cooks are adept at making simple but superb salads of leafy herbs. Together with mint, a large quantity of flat-leaved parsley is a key ingredient in *tabbouleh*, a classic Middle Eastern salad. To the Greeks, a green salad is often a single variety of leafy herb, such as purslane or rocket. In Italy, a plate of rocket scattered with Parmesan shavings and a few drops of olive oil makes a superb starter.

The showier varieties, such as the beautiful purple orache and basils, or the variegated nasturtium leaf, add style to salads. Strew them over a dish of thickly sliced tomatoes (page 38) or use them to bring a supermarket salad to life.

## Nutrition

Eaten in quantity, leafy herbs are a useful source of vitamins and minerals. A couple of handfuls of parsley chopped into a grain- or pulse-based salad, or a big bunch whizzed up into a soup, will provide a day's requirement of vitamin C, nearly 50 per cent of iron and calcium, and over five times the daily requirement of beta-carotene. Basil and rocket provide useful amounts of iron and vitamin C, and, like all green leafy herbs and vegetables, are an excellent source of beta-carotene. Purslane contains useful amounts of potassium and magnesium, while epazote has a high iron content. Most leafy herbs are a good source of calcium.

## Varieties

Of the hundreds of herbs available, we have experimented with dozens over the years, and have found many winners – as well as some we will never grow again. The ones we focused on are the leafier varieties that can be key players in a dish.

### Basil

The basils are a big boisterous group of herbs, with fascinating variations. Their leaves can be small or large, green or purple, glossy or mat, smooth-textured or wrinkled. Essential oils provide a potpourri of flavours and fragrances, including anise, mint, spicy cinnamon and cloves, and citrusy lemons and limes.

Perhaps the best-known and most widely available is **sweet basil** *(Ocimum basilicum)*, with its up-front spicy clove-like flavour. Also easy to find is the small-leaved Greek or **bush basil** *(O. basilicum* var. *minimum)*. It is not quite so intensely spicy as sweet basil, and has a hint of grassiness in the background. With its deep purple serrated leaves, '**Purple Ruffles**' *(O. basilicum)* is really more of an ornamental. It is mildly flavoured, with a hint of anise.

Also attractive is '**Mexican**' basil *(O. basilicum)*, which has a pretty purple stem with pink flowers set off by dark green leaves. It has a strong, complex spiciness backed up by some pungency, and is good in robust bean salads and Mexican dishes.

'**Lettuce-leaved**' basil *(O. basilicum)* has large crinkly leaves with a slightly chewy texture and an assertive flavour that is reminiscent of liquorice and cloves. They are good stuffed with ricotta cheese, chopped toasted pine nuts and sea salt flakes, or diced fried mushrooms and mozzarella mixed with chilli flakes and chives.

'**Lime**' basil *(O. americanum)*, with its citrus flavour, makes a great stuffing or spread chopped and mixed with ricotta cheese, lime zest, mashed prawns and a dash of chilli.

**Thai basil** *(O. basilicum* 'Horapha', not shown) has purple-tinged leaves with a spicy, clove-like aroma. Try it sizzled in hot oil until crisp, with Thai-style grilled prawns (page 106).

### Parsley

Used ubiquitously as a garnish, curly parsley *(Petroselinum crispum)* is one of the most common of the culinary herbs. Nowadays, its flat-leaved cousin, also known as Italian parsley *(P. crispum* var. *neapolitanum)*, is taking the leading role in modern cooking on account of its reputedly better flavour. We tested one simply called **flat parsley**, comparing it with the curly '**Bravour**' (not shown), and found we much preferred the texture and strong clean flavour of flat parsley. It is perfect for the finely chopped mix of parsley, lemon zest and garlic known as *gremolata*. Showered in copious amounts over a rich stew, it lifts and lightens the flavours.

We were curious about some of the parsley look-alikes and tried one called '**Zwolsche Krul**' *(Apium graveolens* Secalinum Group*)*. The leaves and stems have a strong, celery flavour, sometimes to the point of bitterness. Used sparingly, they add flavour to salads, perhaps with a few freshly shelled walnuts and a walnut oil dressing.

**Mitsuba** *(Cryptotaenia japonica)*, also called Japanese parsley, has a flavour that takes some getting used to. It is clean, astringent and mouthwash-like, and we use it cautiously, usually as a garnish. It would probably be good for spiking up rice or grain dishes.

### Sorrel

One of the most acidic of culinary herbs, sorrel has a palate-tingling sourness that is due to the presence of oxalic acid and vitamin C. The somewhat startling flavour can be put to good use in a number of dishes. A few shredded leaves add sparkle to salads, for example, or can make a plain omelette seem like a feast. For the most perfect lemony sauce to accompany salmon, put a large handful of well-washed leaves (no need to chop them) in a pan over gentle heat without any extra water. The leaves will disintegrate to a khaki sludge, which is miraculously transformed by the addition of 300 ml/½ pint of whipping cream and a few spoonfuls of juices from the cooked fish. Heat through and season as necessary.

**Garden sorrel** (*Rumex acetosa*, not shown) has lance-shaped bright green leaves which should be used while still young. We prefer the more mildly flavoured **buckler-leaved sorrel** *(R. scutatus)*. The light green shield-shaped leaves are slightly brittle, with a fresh lemony flavour.

### Purslane

Popular with Greek and Middle Eastern cooks, purslane *(Portulaca oleracea)* is easy to grow. The young leaves have a crunchy texture and pleasant peppery flavour. Dressed with plenty of lemon juice and a little oil, they make a good salad in their own right. For a contrast of colour and flavour, try it with young beetroot, dressed with olive oil and lemon juice.

**Green purslane** has succulent round stems and bright green, slightly mucilaginous, spatula-shaped leaves. There is also a **golden purslane** (not shown), with attractive yellow leaves and bright-red stems which add a certain stylishness to salads.

### Rocket

Rocket (*Eruca vesicaria* subsp. *sativa*), also known as *roquette* and *arugula*, is synonymous with Mediterranean cuisine. It is *de rigueur* in smart designer salads and is an essential ingredient of *mesclun*, the Provençal mixture of tiny salad leaves. Lightly cook the leaves for use in egg dishes or soups or toss them with pasta and olive oil.

We tried two varieties: '**Astro**', with relatively large leaves characterized by shallow notching, and '**Sylvetta**' *(Diplotaxis tenuifolia)*, a so-called wild type with smaller, deeply lobed leaves. 'Sylvetta' has a concentrated spicy flavour with a touch of pungency. 'Astro' is only mildly spicy but more pungent. On flavour alone we would choose 'Sylvetta', though yields may be lower.

### Nasturtium

One of the most striking of herbs, the nasturtium (*Tropaeolum majus*) has vivid orange, yellow or red flowers and saucer-shaped grass-green leaves. We like '**Jewel of Africa**', with its lime-yellow-splashed red leaves. Both leaves and flowers add colour and a refreshing touch of sharp heat to salads. The flavour of the leaves is similar to horseradish, with some lemon in the background. They are delicious rolled around a cream cheese stuffing (page 104).

### Epazote

Given the popularity of Mexican cooking, it is surprising that epazote (*Chenopodium ambrosioides*) remains a stranger to many cooks. Part of the problem may be its aroma, which has been likened to kerosene. However, once cooked in a stew or sprinkled over vegetables, the herb takes on a different quality. It has a clean aftertaste and a slightly chewy texture. The leaves are best used young, while they are tender and relatively mild flavoured.

### Orache

Orache *(Atriplex hortensis)* is used raw in salads or cooked in the same ways as young spinach. Its elongated triangular leaves, with their colour, shape and hint of lemony sharpness, can rescue an otherwise dull salad.

We grew three types: **purple**, **gold** (not shown) and **green** (not shown). Of these, we preferred the mild spinach-like flavour of the green, and the slightly sharp lemon flavour of the yellow. The purple, for all its rich colour, had the least flavour.

Purple orache
Buckler-leaved sorrel
Nasturtium 'Jewel of Africa'
Epazote
Rocket 'Sylvetta'
Green purslane

## Salad of Leafy Herbs ▸

**Serves 4 as a light starter**

*2 handfuls of mixed darkish green leaves, such as red oak lettuce, rocket, baby mustard greens, mizuna, lamb's lettuce and baby Swiss chard*

*4 handfuls of 'Little Gem' leaves (use the pale yellow inner leaves)*

*10–12 very small leaves from the hearts of* cavolo nero *or baby kale*

*10–12 purslane leaves*

*12 small variegated nasturtium leaves*

*16 orange cherry tomatoes*

*4 nasturtium flowers*

**For the dressing:**

*1½ tablespoons cider vinegar or white wine vinegar*

*½ teaspoon Dijon mustard*

*sea salt and freshly ground black pepper*

*6 tablespoons extra-virgin olive oil*

**Some of the quantities may sound overly precise, but they are given simply to show how to create a sparkling salad. Let your imagination and herb garden or market guide you. The idea is to contrast texture, shape and colour.**

1 Whisk the dressing ingredients until smooth and leave to stand for 10 minutes to allow the flavours to blend.

2 Arrange the mixed dark green leaves around the edges of individual serving plates. Place the yellow 'Little Gem' leaves in the middle. Dot with the remaining leaves and the cherry tomatoes. Place a nasturtium flower to one side.

3 Spoon over a little of the dressing – don't use too much or you will drown the leaves – and serve immediately.

## Nasturtium Roll-Ups

**Each filling is enough for 8 leaves**

*24 nasturtium leaves, about 7.5 cm/3 inches in diameter (the variegated variety looks the prettiest)*

*nasturtium flowers, to garnish*

**For filling 1:**

*8 tablespoons spreadable goat's cheese*

*5 nasturtium flowers, chopped*

*3 tablespoons snipped garlic chives or ordinary chives*

**For filling 2:**

*8 tablespoons cream cheese*

*crushed seeds from 3 cardamom pods*

*a little freshly ground black pepper*

**For filling 3:**

*8 tablespoons cream cheese*

*15 g/½ oz smoked salmon, chopped*

*pinch of finely grated lemon zest*

*freshly ground black pepper*

**Nasturtium leaves look good, taste good and roll up nicely. The leaves are free of tough stalks and veins, and they don't split or tear easily. Serve the roll-ups with chilled white wine on a summer evening.**

1 Trim the stalks from the nasturtium leaves at the point where they meet the leaves.

2 Mix your chosen filling with a fork and place about a level tablespoonful on the edge of each leaf. Carefully roll into a cigarette shape, pressing the filling into shape as you roll. Trim the ends neatly with scissors.

3 Arrange the rolls on a serving plate and garnish with nasturtium flowers.

## Parsley, 'Sungold' & Bulgur Salad▸

**Serves 4-6**

*625 g/1 lb 6 oz 'Sungold' tomatoes, halved*
*4 spring onions, thinly sliced*
*75 g/2¾ oz bulgur wheat, rinsed and dried*
*½ teaspoon allspice berries, crushed*
*¼ teaspoon ground cinnamon*
*¼ teaspoon freshly ground black pepper*
*sea salt flakes*
*large bunch of flat-leaved parsley*
*1–2 tablespoons olive oil*
*lemon juice, to taste*
*warm pitta bread, to serve*

**This is a variation on the classic Middle Eastern tabbouleh. We've used a comparatively small amount of bulgur wheat and left out the mint. Parsley and delicious little orange cherry tomatoes are the key ingredients. There is no need to pre-soak the bulgur wheat — the juice from the tomatoes will do the job.**

**1** Put the prepared tomatoes in a bowl with the spring onions and the bulgur wheat. Stir in the allspice, cinnamon, pepper and sea salt to taste. Leave to stand for 2–3 hours or overnight in a cool place – the bulgur wheat will swell as it soaks up the juices.

**2** Strip the parsley leaves from the stems (you should end up with just under 40 g/1½ oz of leaves). Chop them with a very sharp knife rather than a food processor (which would bruise the leaves). Stir into the tomatoes along with the olive oil and lemon juice.

**3** Serve with warm pitta bread.

## Grilled Tiger Prawns with Sizzled Thai Basil

**Serves 2 as a light meal**

*About 24 shelled uncooked headless tiger prawns*
*2 large handfuls of mixed dark green and light green oriental leaves, such as baby bok choy, mustard greens, mizuna, shredded Chinese leaves*
*55 g/2 oz mooli (white radish), cut into matchsticks*
*55 g/2 oz peanuts, toasted and roughly chopped*
*4 tablespoons macadamia oil or extra-virgin sunflower oil*
*6 tablespoons shredded Thai basil*
*freshly cooked jasmine rice, to serve*

**For the marinade:**

*1 large garlic clove, crushed*
*1 lemon grass stalk, tough outer leaves removed, inner stalk very finely chopped*
*1 tablespoon shredded Thai basil*
*1 tablespoon lime juice*
*½ teaspoon sea salt flakes*
*freshly ground black pepper*
*2 tablespoons macadamia oil or extra-virgin sunflower oil*

**For the dressing:**

*1 large garlic clove*
*½ teaspoon black peppercorns*
*1 teaspoon sugar*
*2 tablespoons lime juice*
*½ teaspoon nam pla fish sauce*

**Thai basil has a haunting spicy fragrance. If you can't find any, use ordinary basil instead, and if macadamia oil or extra-virgin sunflower oil prove elusive, use a light olive oil rather than extra-virgin, which is too overpowering for this dish.**

**1** Toss the prawns with the marinade ingredients and leave to stand for at least 30 minutes.

**2** Preheat a grill until very hot. Prepare the dressing: using a pestle and mortar, crush the garlic with the peppercorns and sugar, and combine with the lime juice and fish sauce.

**3** When you're ready to serve, arrange the leaves and the mooli in a shallow serving dish. Toss with the dressing and half the peanuts.

**4** Thread the prawns on skewers. Place under the preheated grill for 5 minutes, turning, until no longer opaque. Remove from the skewers and arrange on top of the greenery.

**5** Heat the oil in a wok until almost smoking. Add the basil and sizzle for 30 seconds. Pour this over the prawns, sprinkle with the remaining peanuts and serve at once, with rice.

# BITTER LEAVES

Though too much can be unbearable, the right amount of bitterness provides vegetables with a much appreciated 'bite'. With their palate-provoking flavours, and offering an impressive diversity that is hard to beat, chicories and endives rank high on our list of worthwhile bitter vegetables.

Chicories *(Cichorium intybus)* are perennials with light green to dark ruby red leaves that are often spectacularly variegated. They range from non-heading types with narrow, deeply-notched, upright leaves reminiscent of dandelions, all the way to the large, red, cabbage-like radicchios. Unlikely as it may seem, the conical white Witloof chicories, with their anaemic scale-like leaves, also belong to this group.

Endives *(C. endivia)* are green-leaved annuals or biennials that form loose heads. There are two basic types: the escaroles or batavias with broad, creased leaves, and the endives or frisées with deeply cut, curly leaves.

## Cultivation and harvesting

The chemicals responsible for the 'bite' of chicories and endives go under the ponderous name of sesquiterpene lactones. Less daunting than they sound, they produce a bitterness that is easily tamed by blanching. With chicories and endives, this simply

Endive 'Elysée'

Chicory 'Catalogna Special'

Endive 'Atria'

Chicory 'Variegata di Sottomarina Precoce'

means that light is excluded from the centre of the plant – usually by tying up the leaves or using covers – so that it turns ivory or white, or, in the case of red varieties, a light pink.

Witloof and some of the red chicories are often 'forced' into early growth by digging up their roots in the autumn and trimming the leaves. The roots are replanted in covered pots and left in a warm spot for three or four weeks until new leaves emerge.

Since they are hardy plants, we grow chicories and endives as a winter crop, both outdoors and in unheated tunnels. Instead of blanching them we grow less bitter varieties, and we either harvest the whole head at one go or pick the outer leaves as they develop.

## Buying and storing

Look for radicchio with tightly packed firm heads and reject tired specimens past their prime. Elongated or conical chicories should not have slimy brown patches along the leaf edges. Make sure that frisée and escarole have big, sturdy blanched hearts and not too many dark green outer leaves. Though somewhat expensive to buy, chicories and endives represent good value for money. A few leaves go a long way and they store for at least a week in the fridge.

## Preparation

Prepare loose-leaved chicories and endives as you would ordinary lettuce. To prepare Witloof chicory, remove the outer leaves, trim the root end, then separate into single leaves.

## Cooking

Chicories and endives are unexpectedly versatile. Not only are they superb salad material, but they can also be grilled, oven-baked, sautéed, braised or added to risottos and pasta dishes. For a simple and tasty pasta sauce, stir-fry roughly torn leaves of escarole or red chicory in olive oil until barely wilted, then toss with pasta, Parmesan shavings and basil, and sprinkle with toasted bread-crumbs. Alternatively, stir the cooked leaves into a creamy risotto.

Bitter-tasting leaves are an excellent foil to rich meat dishes such as grilled venison fillet or Balsamic Vinegar-Glazed Rabbit (page 110). We like to serve the leaves with the meat, but a large salad, perhaps dressed with walnut oil and balsamic vinegar, is equally delicious after the main course.

Witloof chicory is good halved lengthwise, generously brushed with oil and blasted under a very hot grill. The edges of the leaves will be slightly charred and the hearts tender but crisp. Sprinkle with sea salt and plenty of chopped parsley before serving. It is also good braised with butter, sugar and lemon juice, then sizzled on top of the stove until richly glazed.

## Nutrition

Chicories and endives contain insignificant amounts of most nutrients. However, they do provide massive amounts of carotenes – not only beta-carotene, but also lutein and zeaxanthin.

## Varieties

Our tastings were carried out in early winter on a range of chicories and endives planted outside in late summer. The plants were unblanched, and only the centre leaves were tasted.

### Chicories

Too bitter for us were the green '**Bianca di Milano**' (not shown) and the red '**Variegata di Sottomarina Precoce**'. Another one that should be used with some discretion is '**Catalogna Special**', a tall dandelion-leaved type. However, its bitterness was reined in somewhat by lightly cooking it and serving it in a warm salad with balsamic vinegar (page 110). A milder variety is '**Variegata di Chioggia**' (not shown) with red-splashed leaves, while '**Variegata di Castelfranco**' (not shown) is the least bitter of the group.

We also tried two red chicories supplied by a specialist importer. The long, tapering '**Rosso di Treviso**' (not shown) is a remarkably handsome plant, with white-veined burgundy-coloured leaves that were not overly bitter. It makes a stunning salad with yellow-fleshed ocas and purple potatoes (page 165). '**Tardivo di Treviso**' (not shown) is a small, late-maturing variety, prized for its elegant leaves, crunchy texture and sweetish flavour.

### Endives

We liked the sweetness and texture of frizzy-leaved '**Sally**' (not shown), the broad-leaved '**Jeti**' (not shown) and '**Elysée**'. Our favourite, though, was the frizzy '**Atria**', with a nice crisp texture, pleasant sweetness and an assertive though mild bitterness.

## Balsamic Vinegar-Glazed Rabbit with Bitter Leaves▸

**Serves 6-8**

*1.3 kg/3 lb rabbit joints, such as loin cutlets and shoulder portions*
*150 ml/¼ pint balsamic vinegar*
*25 g/1 oz butter*
*2 tablespoons extra virgin olive oil*
*450 g/1 lb baby onions or shallots, peeled and left whole*
*140 g/5 oz shiitake mushrooms, sliced*
*2–3 garlic cloves, finely chopped*
*salt and freshly ground black pepper*
*1 tablespoon chopped fresh thyme*
*600 ml/1 pint chicken or meat stock, preferably home-made*
*2 tablespoons chopped flat-leaved parsley*
*6–8 good handfuls mixed bitter leaves*
*garlic croûtons (page 126), to garnish*

**This is a Tuscan dish of which the leaves are an integral part rather than merely a garnish. Their slight bitterness is perfectly balanced by the sweetness of balsamic vinegar. Use the beautiful green and red spotted 'Variegata di Chioggia', radicchio or even Witloof. Make sure the leaves are thoroughly dry. Supermarkets sell packs of rabbit joints which are ideal for this dish.**

**1** Put the rabbit joints in a frying pan large enough to take them in a single layer. Pour in the vinegar and simmer over moderate to high heat, turning several times, for about 15 minutes, until the vinegar has turned into a deliciously sticky coating to the rabbit. The pan will be quite dry towards the end. Remove the rabbit from the pan.

**2** Heat the butter and oil in the same pan. When the butter is foaming, throw in the onions and mushrooms and add 5–6 tablespoons of the stock. Stir everything around so that the vegetables are coated with the stickiness. Next, add the garlic, cook for a minute, then pour in a little more stock to moisten.

**3** Return the rabbit to the pan and pour in about 300 ml/½ pint of the stock. Season with salt and pepper and add the thyme. Cover and leave to cook for 25–30 minutes, until the rabbit and onions are tender. Sprinkle with the parsley and check the seasoning.

**4** Arrange the leaves on a large platter or on individual serving plates. Place the rabbit pieces, onions and mushrooms on top and garnish with croûtons. Add the remaining stock to the pan and bubble down the juices over high heat. Pour over the rabbit and serve at once.

## Wilted Bitter Salad with Bacon

**Serves 2 as a starter**

*85 g/3 oz finely diced bacon lardons*
*3 tablespoons extra-virgin olive oil*
*2 garlic cloves, thinly sliced*
*2 large handfuls trimmed young dandelion-leaved chicory or pissenlit*
*salt and freshly ground black pepper*
*2 teaspoons balsamic vinegar*
*2 freshly cooked hard-boiled eggs, halved*
*warm ciabatta bread, to serve*

**Use a dandelion-leaved chicory such as 'Catalogna Special' or young dandelion leaves, sometimes sold as 'pissenlit' in the shops. If you use wild dandelions, make sure they are gathered from a traffic- and dog-free area. The relative sweetness of bacon and balsamic vinegar counteracts the bitterness of the leaves.**

**1** Using a small non-stick frying pan, fry the bacon lardons without any oil until crisp.

**2** Heat the olive oil in a large frying pan over moderate heat. Add the garlic and gently fry until just beginning to colour. Throw in the leaves, season with salt and pepper, and stir for a few minutes until wilted. Splash with the balsamic vinegar.

**3** Divide between two plates, pour over the juices and top with crisp lardons and the hard-boiled egg halves. Serve immediately with plenty of ciabatta bread.

'Menuet'
CABBAGES

# CABBAGES & KALES

Cabbages *(Brassica oleracea* Capitata Group*)* and kales (see individual varieties) are among the sturdiest and most benevolent of vegetables. They defiantly stand up to winter frosts and, in the hands of a loving cook, they feed you generously and well. Both are brassicas, though judging by their external appearance it is hard to see just how they are related. Closer examination reveals that they have much in common.

Kales are non-heading vegetables with long graceful leaves growing from a central stalk. In contrast, most cabbage varieties develop heads which grow close to the ground. Their heads, however, are something of an optical illusion; it is only when one is sliced from top to bottom that the true nature of the cabbage is revealed. It is, in fact, a very dwarf plant with closely packed leaves lining a central stalk. The set-up is similar to the kale's, except that in the cabbage the cup-like lower leaves enclose those further up in a neat Russian doll arrangement that forms the head.

'January King-3'

## A culinary cornucopia

Unassuming and often unappreciated, both cabbage and kale offer a culinary cornucopia of colours, textures and flavours. Kale may be smooth or blistered, deeply toothed or flamboyantly frilled. Though most varieties come in a basic shade of green, some break the mould with splashes of red, while others are a mysterious blue-green so dark it could almost be black. Cabbages can be crinkled like the Savoy, or smoothly etched with intricate veining. They have a rich palette of colours that includes deep emerald greens, royal purples, the palest of creams, and luminous blue-greys. Flavours are as diverse as varieties, ranging from the mildness of white cabbage to the meatiness of some types of kale.

## Cultivation and harvesting

Chemicals known as glucosinolates are responsible for the sulphurous flavour and pervasive cooking smell of cabbage and kale. To a certain extent, gardeners can manipulate glucosinolate levels by juggling the right variety with different horticultural practices. Levels are increased by close planting distances, water stress and high concentrations of sulphur in the soil, so attempting to influence flavour becomes a gardening game of multiple choice.

Kale is traditionally a winter crop; cabbage, with a careful selection of varieties, can be harvested throughout the year. Even so, we prefer the heading cabbages that are ready either side of Christmas and which, along with the kales, provide good eating in an otherwise lean season. Both have a long growing season and need to be sown in summer and lovingly nurtured through the autumn.

## Buying and storing

If possible, choose cabbages with the outer leaves still attached since these not only help keep the cabbage fresh but also contain more nutrients. The leaves should show no sign of limpness or yellowing, and the hearts should feel heavy and solid. Kale leaves should be full of bounce, again with no sign of yellowing. Examine packets of baby kale and reject any that look bruised or slimy.

If left uncut and loosely wrapped in plastic, a firm fresh cabbage can be stored in the fridge for a week or two, although there may be some moisture loss. Kales deteriorate much more quickly and will keep for only 2–3 days before turning yellow.

## Preparation

With nutrient levels in mind, try to use as many dark green outer cabbage leaves as possible. Unless very leathery or damaged, they are perfectly edible sliced into ribbons with the stalks removed.

For wedges, or for shredded cabbage, quarter the head lengthwise, then cut out the core. To shred the leaves, cut the quarters across into thin ribbons. Don't discard the core: it is excellent chopped into salads, or trimmed neatly and served as a crudité with some good mayonnaise or a thick lemony vinaigrette.

To remove tough stalks from kale, hold a folded leaf in one hand, grasp the stalk with the other and strip it away. Then chop the leaves according to the recipe.

## Cooking

The basic rule with cabbages and kales is short cooking time – 5–7 minutes at most – for crispness and colour, or very long cooking to bring out the sweetness; anything in between is a disaster.

There are two schools of thought about boiling: one is to use a large volume of ready-boiling water, the advantage being speed, and therefore the retention of colour, texture and flavour. The second way is to use a minimum amount and thus retain more of the vitamins. We go for the first method; or we stir-fry them.

Cabbages and kales should be cooked uncovered. In a covered pan, there is a reaction between the green chlorophyll pigments and plant acids that would otherwise escape with the steam, resulting in a drab olive green colour. To keep red cabbage bright, add a little vinegar or lemon juice to the cooking liquid.

Kale does not lend itself to anything more complicated than boiling or steaming, although it can be used to good effect in a soup if not allowed to dominate. Baby kale is delicious in stir-fries, or used raw in salads dressed with a little walnut oil.

'Red Russian'
'Hungry Gap'
KALES
'Black Tuscany'
'Showbor'

Cabbage is far more versatile than kale. It can be served raw or plainly boiled, topped with melted butter or cream or stir-fried with spices, stuffed with any number of fillings or slowly simmered in a soup or casserole. It also forms the basis of the fermented pickle sauerkraut.

Cabbage features prominently in the cooking of Eastern Europe and Russia, usually slow-cooked and often seasoned with paprika or dill. It is a key ingredient in *shchi*, a hefty Russian soup of fresh or fermented cabbage and boiled meats, and in *garbure*, a French soup which may include potatoes, beans, pork or goose.

Slowly cooked with onion, apples and port, red cabbage is the classic accompaniment to roast goose or pork. It is also good stir-fried, although you may need to blanch it first to rid it of some of its bounce. A few cranberries thrown in add brilliant nuggets of colour. Or try it stuffed with a mixture of chopped crisp nuts, wild rice, lemon zest and garlic (page 117) – wonderful served with a zinging grilled tomato or roasted red pepper sauce.

Coarsely shredded white cabbage is brilliant for salads. It adds bulk and texture, and does not wilt even when confronted with a strong acidic dressing. We like to marinate it in cider vinegar and sugar with chillies and red onion, then after a day or two we add some waxy new potatoes and a little oil (page 117).

## Nutrition

Cabbages and kales contain a whole string of cancer-fighting phytochemicals which include indoles, chlorophyll, phenols and flavonoids, as well as the ubiquitous glucosinolates. They also contain vast amounts of carotenes, and are rich in vitamins C and E – all antioxidants known to help combat cancer.

Recent research the world over provides convincing evidence of a link between regular cabbage consumption and reduced rates of colon cancer. Cabbage is also believed to speed up the metabolism of oestrogen, which may help protect against breast and ovarian cancer. Kale is linked with lower incidences of colon, prostate and bladder cancer.

In addition to their cancer-fighting properties, cabbages and kales are a rich source of potassium and folate, and provide useful amounts of calcium, thiamin (vitamin $B_1$) and pyridoxine (vitamin $B_6$).

## Varieties

Our tasting trials were carried out in early winter. Wedges of cabbage and the upper leaves of the kales were steamed. The cabbages were also tasted raw.

### Cabbages

The outer leaves of '**January King-Hardy Late Stock 3**' are a rich dark green overlaid with bright red – a dramatic combination which unfortunately does not extend to the heart. It is well flavoured with a sweetness which really comes across with long, slow cooking in a rustic soup or stew.

Another worthwhile variety is '**Menuet**', a Savoy type with a soft head and deeply blistered dark blue-green leaves. It has a pleasant touch of pungency, and a full satisfying 'cabbage' flavour. Still good but not as flavourful is '**Tundra**' (not shown), a Savoy and white cabbage hybrid with light green, slightly blistered leaves that are crisp and 'squeaky' when cooked. The leaf stalks have a sweetness that is offset by a mustardy flavour. 'Tundra' is excellent stir-fried in extra-virgin sunflower oil with crushed juniper berries, lemon zest and flat-leaved parsley. The red '**Autoro**' has a pleasant mix of pungency and sweetness, and is good stuffed or stir-fried.

None of the green or red varieties did well in our raw taste tests – they were too dry and chewy. However, a white variety, '**Apex**' (not shown) has a crisp, juicy texture, and a distinctly sweet and subtly pungent flavour. It makes a good coleslaw, and is excellent shredded in a salad with carrots, kohlrabi and pumpkin seeds.

### Kales

'**Showbor**' *(B. oleracea* Acephala Group*)*, a short-statured, curly-leaved variety, has tasteless leaves and tough stalks. So too does '**Hungry Gap**' *(B. napus* Apisera Group*)*, though its leaf stalks have a sweet, almost minty, kale flavour. We found that '**Black Tuscany**' *(B. oleracea* Acephala Group), a type of *cavolo nero*, tasted bland; what flavour there was had migrated to the very chewy stalks.

The kales were redeemed by '**Red Russian**' *(B. napus* Pabularia Group*)*, a smooth-leaved variety with a touch of red. Though the stalks were stringy and flavourless, the blades had a soft texture and superb flavour. It was delicious stir-fried with ginger, soy sauce and sesame seeds.

## ◂Red Cabbage with Wild Rice, Mushroom & Nut Stuffing

**Serves 4–6 as a main course**

*150 ml/¼ pint wine vinegar or light malt vinegar*
*1 red cabbage, weighing about 1 kg/2¼ lb*
*175 g/6 oz mixed nuts, such as Brazil nuts, cashews, hazelnuts and almonds*
*2 tablespoons olive oil*
*40 g/1½ oz butter*
*1 onion, finely chopped*
*2 teaspoons chopped fresh thyme*
*1 teaspoon chopped fresh rosemary*
*140 g/5 oz mushrooms, chopped*
*115 g/4 oz cooked wild rice*
*3 garlic cloves, finely chopped*
*3 tablespoons chopped flat-leaved parsley, plus some whole leaves to garnish*
*finely grated zest of 1 lemon*
*½ teaspoon salt*
*freshly ground black pepper*
*300 ml/½ pint stock*

**It is normally green cabbage that gets stuffed. Here we give red a chance. Adding vinegar to the cooking water stops the cabbage from turning blue.**

**1** Preheat the oven to 180°C/350° F/gas 4. Bring a very large saucepan of water to the boil and add the vinegar. Cut about 4 cm/1½ inches from the base of the cabbage. Carefully peel away 6–8 outer leaves, trying not to tear them too much. Plunge them into the boiling water for 3–4 minutes. Drain under cold running water and pat dry. Using a small, sharp knife, shave away the thick part of the stalk so that the leaves are more pliable.

**2** Spread the nuts out on a baking sheet and roast in the oven for 5–8 minutes, until golden. Chop them by hand – if you use a food processor the nuts will turn to dust.

**3** Cut the remaining cabbage in two and reserve one half for use in another recipe. Slice the other half lengthwise into four. Cut out the core, and cut the leaves across into thin shreds.

**4** Heat the oil and 25 g/1 oz of the butter in a large frying pan. Add the onion, thyme and rosemary, and gently fry for a few minutes until the onion is translucent. Throw in the mushrooms and the shredded cabbage and fry for 5 minutes. Stir in the rice, nuts and garlic, and fry for 2 minutes more.

**5** Remove from the heat and stir in the parsley, lemon zest, salt, a generous amount of freshly ground pepper and 150 ml/¼ pint of the stock.

**6** Lightly grease an 850 ml/1½ pint round ovenproof dish, about 22 cm/8½ inches in diameter. Arrange the blanched cabbage leaves around the edge, overlapping them so that there are no gaps. Pile the stuffing in the centre, packing it in well. Dot with the remaining butter and fold over the tops of the leaves. Pour the remaining stock round the outside of the leaves.

**7** Tightly cover the dish with a double thickness of foil and bake at 180°C/350° F/gas 4 for 45 minutes. Serve cut into wedges, and accompanied by the tomato sauce.

## Sweet and Sour Cabbage & Potato Salad

**Serves 4–6**

*2 teaspoons sugar*
*½ teaspoon salt*
*150 ml/¼ pint cider vinegar*
*½ small white cabbage, thickly sliced*
*1–2 fresh red chillies, deseeded and thinly sliced*
*½ red onion, thinly sliced*
*350 g/12 oz cooked new potatoes, thickly sliced*
*freshly ground black pepper*
*4 tablespoons extra-virgin sunflower oil*
*2 tablespoons chopped fresh coriander*

**You can eat this straight away but it is better to let it stand for a day or two. The onion tinges the cabbage a beautiful shade of pink.**

**1** Dissolve the sugar and salt in the vinegar. Mix with the cabbage, chillies and onion in a non-reactive bowl. Cover and leave to marinate in the fridge for 1–2 days, stirring occasionally.

**2** Mix in the potatoes, season, and toss with the oil and coriander.

## New Mexican Caldo Verde▸

**Serves 6**

*4 tablespoons groundnut oil*
*500 g/1 lb 2 oz lean boneless pork, cut into bite-sized pieces*
*2 white onions, thinly sliced*
*1 teaspoon cumin seeds, crushed*
*½ teaspoon dried oregano*
*1 litre/1¾ pints chicken stock, preferably home-made*
*2 large garlic cloves, very finely chopped*
*1–3 jalapeño chillies, deseeded and finely chopped*
*3 tomatoes, peeled and chopped*
*350 g/12 oz cooked appaloosa beans, borlotti beans or butter beans*
*½ teaspoon salt*
*½ teaspoon freshly ground black pepper*
*140 g/5 oz trimmed and shredded kale*
*5 tablespoons chopped fresh coriander*
*1–2 tablespoons lime juice*
*corn tortillas, to serve*

**To garnish:**
*thinly sliced white onion*
*diced avocado, tossed in a little lime juice*
*lime wedges*

**This is a great main meal soup for a casual weekend dinner with friends. It is based on the classic Portuguese soup of pork, beans and greens, but given a New Mexican twist with avocado, coriander and lime juice. Use lightly cooked fresh shelling beans, or dried beans that have been soaked and boiled for 20 minutes, or even drained canned beans if you are feeling lazy.**

**1** Heat the oil in a large saucepan over moderate to-high heat. Add the pork and fry until lightly browned. Remove from the pan with a slotted spoon and set aside.

**2** Reduce the heat to low and add the onion, cumin and oregano, stirring to coat them in the fat. Cover and cook for 25–30 minutes, stirring occasionally, to prevent sticking. Add a splash of stock if necessary. When the onion is very soft, add the garlic, chillies and tomatoes. Cover and cook for 10 minutes.

**3** Next, add the meat, beans, stock, and salt and pepper. Bring to the boil, then reduce the heat and simmer gently, with the lid slightly askew, for 45 minutes, until the meat is tender.

**4** Raise the heat to a brisk simmer and stir in the kale. Cover and simmer for 5–7 minutes more, until the kale is tender but still bright green. Stir in the coriander and lime juice just before serving. Serve with the garnishes and tortillas.

## Greens & Beans with Chorizo

**Serves 4 as a main course**

*3 tablespoons olive oil*
*1 large white onion, thinly sliced into crescents*
*3 garlic cloves, finely chopped*
*2–3 large mild green chillies, such as Anaheim, deseeded and roughly chopped*
*1 teaspoon dried oregano or thyme*
*175 g/6 oz chorizo sausage, thickly sliced*
*250 g/9 oz shredded Savoy cabbage*
*300 ml/½ pint chicken stock, preferably home-made*
*280 g/10 oz cooked butter beans or other large beans*
*salt and freshly ground black pepper*
*3 tablespoons chopped fresh coriander*

**Served with hunks of good bread, this makes a fortifying lunch or supper dish. You can use fresh shelling beans instead of dried if you have them.**

**1** Heat the oil in a large frying pan with a lid and fry the onion over medium heat until soft but not coloured. Add the garlic, chillies and oregano or thyme, and fry for another minute or two. Then stir in the chorizo and let it sizzle for a bit.

**2** Add the cabbage and stock and bring to the boil. Mix in the beans and season to taste with salt and pepper, bearing in mind that chorizo is quite highly seasoned.

**3** Cover and simmer for 5–10 minutes, depending on how tender you like your cabbage. Sprinkle with coriander before serving.

# BULBS, ROOTS & TUBERS

ALLIUMS BEETROOTS SWEDES, TURNIPS & RADISHES CELERIAC & KOHLRABI

FENNEL CARROTS JERUSALEM & CHINESE ARTICHOKES

SALSIFY, SCORZONERA & HAMBURG PARSLEY OCA & MASHUA

JICAMA & BURDOCK POTATOES

# ALLIUMS

Alliums have split personalities – slice an onion and it will make you cry, crush some garlic and you'll be struck by the smell; left intact, though, both are innocuous. The secret of this Jekyll and Hyde behaviour lies in the sulphurous chemical precursors found in the plants' cells. When the cells are damaged by cutting or crushing, the precursors interact with an enzyme known as alliinase, and are transformed into the volatile chemicals responsible for the typically potent flavours and odours.

The potency of an allium dish depends on how the alliums are prepared and cooked. If, for example, garlic is crushed rather than chopped, more cells are damaged, more volatiles are created, and industrial-strength flavours and odours begin to dominate.

The growing environment and physiological state of the plant also leave their mark. High levels of sulphur in the soil tend to pump up the pungency levels in onions, while those deprived of water often pack more flavour than specimens raised under lusher circumstances. Moreover, as if to go out with a bang instead of a whimper, they markedly increase their production of sulphurous volatiles as they begin to sprout.

## Onions

The common onion (*Allium cepa* Cepa Group) comes in a range of shapes, sizes and colours, ranging from bulbous yellow bruisers to streamlined pink torpedoes and pencil-thin salad onions. The varieties most common in the shops and seed catalogues are yellow- or brown-skinned, with creamy flesh and pungency levels ranging from moderate to extreme. 'Spanish' onions are similar in colour, but much larger and juicier, with a milder flavour.

White onions have a pearly paper-thin skin and deliciously crisp, juicy flesh that is excellent raw or cooked. Red onions have attractive purple and white flesh, with a similar texture. Pungency in both can be unpredictable – some are mild, while others are occasionally unexpectedly sharp, pungency often kicking in after an initial impression of sweetness.

So-called 'sweet' onions are often named after their place of origin – Washington's '**Walla Walla Sweet**', for example. Flavour, however, depends on the growing conditions and may not be so exceptional when a crop is grown away from its home town.

Growing bulbing onions is simply a matter of sowing sets or seeds in late winter or early spring. If allowed to mature, the onions will be ready for harvesting in late summer or autumn, when the tops die down and a protective skin forms around the bulbs. In areas with mild climates, some varieties can be sown in the autumn and grown through the winter. These mature sooner than spring-sown varieties, and are useful when early onions are needed. Salad onions will produce small bulbs if left in the ground long enough. The rule for sowing them is 'little and often'. With a careful selection of varieties, they can be grown throughout the year in milder areas.

However, not all onions produce bulbs. The so-called Japanese bunching onion, often grown as a salad crop, the Welsh onion (both *A. fistulosum*) and the everlasting or ever-ready onion (*A. cepa* var. *perutile*) remain virtually bulbless even when mature. All three are perennials and produce numerous stems in tight clumps that can be split for propagation. The Welsh and Japanese onions can also be grown from seed, which is the preferred method for the Japanese types when they are grown as a single-stem salad crop.

## Shallots

Shallots (*A. cepa* Aggregatum Group) have a distinctive flavour of their own. Traditionally started from sets, they can now be grown from hybrid seeds. Sown close together, the hybrids produce small single bulbs, though we have managed so far to grow only large specimens, which sometimes split into clusters joined at the base.

'Long Red Florence'

ONIONS

'Red Baron'

'Matador'

SHALLOTS

'Owa'

'Creation'

'Walla Walla Sweet'

'Red Baron'

'Albion'

'Paris Silverskin'

## Leeks

Depending on the variety and the time of harvesting, leeks *(A. ampeloprasum* Porrum Group*)* range from slim willow-green babies to sturdy fat-shanked monsters. Cultivation normally involves planting into a hole (dibbing) in the ground and earthing up two or three times to blanch the stems and induce what botanists call 'etiolation'. In non-technical terms, this simply means that the subterranean darkness encourages the stems to grow longer and whiter, at the same time becoming more succulent and tender, as the plants develop.

To produce baby leeks, the seeds should be sown close together to crowd the plants and keep them small. With a sufficiently early sowing, harvesting of the young plants can start by mid-summer.

## Garlic and garlic-flavoured alliums

There is more to garlic than meets the nose – it's a family with some well-kept secrets. As well as various strains of 'common' garlic *(A. sativum)*, there is the less familiar elephant garlic *(A. ampeloprasum* Ampeloprasum Group*)*, with unnaturally huge cloves, and the small-cloved rocambole *(A. sativum* var. *ophioscorodon)*, growing contentedly in the wild and in cottage gardens.

In contrast are garlic or Chinese chives *(A. tuberosum)*, which are something of a maverick, since the parts usually eaten are the broad flat leaves rather than their tiny bulbs. Garlic chives are grown from seeds, eventually forming clumps that can be split and replanted. The other types of garlic are propagated by planting cloves either in autumn or spring, though rocambole has diversified somewhat. It produces bulbils at the end of its long spiral stems and these can be used to start off a new crop.

Garlic bulbs may be harvested as soon as they begin to swell. The new season's 'wet' or 'green' garlic, with its moist, pliable skin and barely formed cloves, is a joy. Alternatively, the bulbs can be left in the ground until they reach full size and then pulled and left until the skin dries and forms a protective covering.

## Buying and storing

When buying onions, shallots and garlic, choose plump, solid bulbs with a tight skin. Reject any that are soft or mouldy. Spring onions should look sprightly, without any withering or sliminess, and the white part should be firm. The same applies to leeks.

To prevent rotting and sprouting, bulbing onions and garlic need to be stored in a cool dark airy place. Both will keep for several months under the right conditions. Exceptions are the juicy Spanish and red onions, and the 'sweets', which start to deteriorate after a week or so. Leeks and spring onions should be stored in the salad drawer of the fridge and used within a few days. Trim the tops if they are overly exuberant.

## Preparation

Explaining how to prepare onions is rather like explaining how to boil an egg. However, now and again it is worth going back to basics. It is important to use a sharp knife and to work in an orderly way, since haphazard hacking destroys more cells and increases the likelihood of tears. Slice the peeled bulb in half lengthwise and place it, cut side down, on the work surface. Thinly slice each half again lengthwise or across, depending on whether you want 'wings' or crescents. To chop, slice first in one direction, holding the slices tightly together, then slice neatly at right angles to the previous set of cuts.

Leeks invariably need cleaning. The easiest way is to slice them, swish them in a bowl of water, then drain and dry thoroughly. If you want to use a leek whole, make a cross-cut down the middle, stopping short at the white part, open out the layers and invert it in a bowl of water for 15 minutes.

When preparing garlic, remember that the strongest flavour is produced by crushing the clove with the flat of a knife and mashing the flesh with salt using a pestle and mortar. Cloves can also be finely chopped or, for the mildest flavour, left whole.

## Cooking

Onions form the backbone of most savoury dishes, providing body and unifying flavour. In the majority of recipes, the first step is to soften chopped or sliced onions slowly in oil. This gentle preamble is an essential part of their flavour development and should not be hurried, so try to refrain from adding other vegetables until the onions have cooked down. Garlic, however, can be added once the onions are translucent and soft – it tends to burn and become acrid if you add it earlier.

For stewed onions, fry slices or rings over very gentle heat for 30–40 minutes until the onions melt down to a deliciously soft, mahogany-coloured mass. Add a splash of wine vinegar towards the end to spike up the flavour. Whole onions can be baked alongside a roast; they will caramelize beautifully, adding flavour and rich colour to the juices.

Baby onions or pickling onions are good cooked whole in a rich braise or daube, such as *coq au vin* or the Greek *stifado*. They are also useful for kebabs. Unless they are very small, cut them in half lengthwise before threading them on skewers.

Red onions are delicious sliced into the thinnest of crescents and scattered over salads – try them with blood oranges and fennel; or dice them finely and add to a tangy salsa with nuggets of mango and some chopped coriander – perfect with grilled fish. Since their sugar content encourages them to blacken nicely around the edges, they also make excellent barbecue material.

Finely chopped spring onions are delicious sprinkled over sliced tomatoes or a new potato salad. For a subtly flavoured raita, slice them finely and mix them with yoghurt and crushed cumin seeds. Don't throw away the green tops: sliced into rings or finely chopped, they make a colourful and flavoursome garnish. Spring onions also tolerate light cooking. The Chinese add them to stir-fries, although they are more likely to use the Japanese bunching onion. Try them French-style, too, seethed in butter with peas.

Leeks need centre stage to show off their flavour. They are a key ingredient in big soups, such as cock-a-leekie and the Welsh *cawl*. Combined with eggs or mild cheese, they make a supremely satisfying tart (see page 131). Baby leeks are delicious roasted – they need no more than a lick of oil, a sprinkle of sea salt and a fifteen-minute blast in a very hot oven. We also like sliced leeks shaken for a few minutes in a covered pan with a dab of butter; they cook in their own steam. Tender leeks are also very good uncooked – paper-thin slices add a definite pep to salads.

When it comes to garlic, cooks can be as discreet or upfront as they please. Rubbing the salad bowl with a cut clove leaves just a hint; smearing a crushed clove over toasted bread is more aggressive. Simple and delicious is spaghetti tossed with lightly fried sliced garlic cloves, dried chilli flakes, lots of olive oil and a generous sprinkling of chopped flat-leaved parsley. The result is unmistakably garlicky, but it is tempered by the pasta and the clean-tasting parsley.

Garlic can also be treated as a vegetable in its own right. Slice the tops off large unpeeled heads and roast the heads around a joint of meat, or slice them in half horizontally, brush them with good olive oil and cook them on the barbecue. The flavour is deliciously sweet and creamy – quite unlike fried garlic.

Also delicious is a Three-Way Garlic Soup (page 131), served with garlic cream and croûtons. For garlic cream, simmer four whole peeled cloves with 250 ml/8 fl oz double cream and a sliver of lemon peel, until the cream has thickened. Push through a sieve, season with salt, pepper and a few drops of lemon juice, and swirl into the soup. For garlic croûtons, heat some olive oil with a few slices of garlic. Remove the garlic when it is coloured, then toss cubes of stale bread into the pan and fry until golden. The croûtons will keep in an airtight tin for a week or two.

## Nutrition

Onions and shallots both contain minuscule amounts of potassium, calcium, iron, zinc and B vitamins. Shallots, however, contain over twice the vitamin C of onions. The white parts of salad onions provide a useful amount of calcium – weight for weight nearly six times more than ordinary onions. With the green tops included, they are an excellent source of carotene.

Leeks contain small but significant levels of fibre, potassium, calcium and iron. An average-sized serving provides more than the daily requirement of carotene and a goodly amount of vitamin B6 and vitamin C.

Garlic scores the highest in health-promoting properties. It contains a significant amount of protein, carbohydrate and fibre, and is a useful source of potassium, magnesium, iron and vitamin C. Current research suggests that garlic may be of use in the prevention of coronary heart disease and strokes. To be effective, however, it needs to be eaten raw, regularly and in large amounts.

## Varieties

The simple techniques we normally employ in tasting trials – steaming or light boiling, for example – are chosen so as not to detract from the flavour of the vegetable. With the exception of leeks, this strategy does not apply to alliums, since their ultimate flavour is dependent on more complex cooking techniques and the alchemy that takes place between them and other ingredients. However, comparative testing using fully cooked dishes would not only be impractical but would also introduce other variables. Our tests, therefore, are not necessarily a full indication of the flavours that could potentially develop.

### Onions and shallots

We tested bulbing onions and shallots by frying them in a neutral oil until just soft. Outstanding for their rich flavour were our giant seed-sown shallots '**Matador**' and '**Creation**'. Among the onions we liked the stunning red-skinned '**Red Baron**' and white-skinned '**Albion**'. Cooked, both were sweet and mild. Tasted raw, they were both distinctively sweet, but almost unbearably pungent. Sliced very thinly and used in moderation, though, they added a nice piquancy to a gutsy roast beef salad. '**Owa**', a yellow-skinned elongated type, was not especially sweet but had more flavour than 'Red Baron' or 'Albion'. We liked it roasted with olive oil, rosemary and breadcrumbs (page 128). In contrast, '**Long Red Florence**' was neither flavourful nor particularly sweet.

The thin-stemmed **everlasting onion** was troublesome to peel and cook. It quickly became crisp and fibrous and had little flavour. Used raw, though, it was mildly pungent and sweet. With its thicker stems, '**Red Welsh**' onion was easier to deal with, and had a nice, fairly robust flavour when cooked in a stir-fry.

Of the salad onions we tested, '**Guardsman**' (not shown) and '**North Holland Bloodred-Redmate**' (a variety normally grown for its bulbs) distinguished themselves. Both have tender, slightly sweet and mildly pungent stems. Our favourite variety for eating raw is '**Purplette**', with its crisp, sweet, purple-skinned bulbs. It is also good lightly cooked. Try it with peppers, tomatoes and courgettes in a colourful Mediterranean stir-fry (page 20).

Another bulbing variety we sometimes use raw is '**Buffalo**' (not shown). We like its sweet herbaceous green tops and mildly pungent bulbs that remain sweet and delicious even up to 4–5 cm/1½–2 inches in diameter. We also cook them in a sweet-and-sour dish to serve with pork or a Christmas bird (page 128).

'**Paris Silverskin**' is an older white variety normally used for pickling. We like both its leaves and the immature bulbs, though the mature bulbs are too pungent to eat raw.

The American '**Walla Walla Sweet**' failed to live up to its reputation as a gourmet onion. Our full-size bulbs, though juicy, were overly mild and not especially sweet. Young bulbs had some pungency and sweetness, but the green tops were quite tasteless. It obviously does not travel well.

### Leeks

We are lazy gardeners and so don't blanch our leeks. As a result, the stems are greener and more fibrous, but we have come to prefer the robust texture. We grew '**King Richard**' and '**Varna**' (not shown), both naturally long-stemmed varieties. Harvested in late winter, they were sliced and steamed for taste-testing. Both had a pleasantly mild 'oniony' flavour, though 'Varna' was sweeter and slightly more tender. We like it raw in salads, sliced paper-thin and tossed with tomatoes, black olives and chunks of hard-boiled egg.

### Garlic

Reassured by the potential health benefits, we stoically tasted the garlics raw. **Rocambole** and the 'common' garlics '**Cristo**' and '**Germidour**' were all pungent, with a nice assertive garlic flavour. In contrast, and as expected, **elephant garlic** was much tamer and, to us, less satisfying. It was pleasant enough, however, roasted or cooked in a creamy sauce. Leaves of **garlic chives** had a herbaceous, very mild garlic flavour which became stronger when sprinkled over cooked vegetables. The flowers make an attractive garnish for a clear oriental-style soup (page 98).

## Roasted Torpedo Onions▸

**Serves 4–6 as a side dish**

*8 torpedo-shaped onions, such as 'Owa'*
*olive oil for brushing*
*55 g/2 oz fresh breadcrumbs, made from a stale ciabatta loaf*
*1 tablespoon chopped fresh rosemary*
*sea salt flakes*
*coarsely ground black pepper*
*85 g/3 oz fresh Parmesan cheese*

**Even if you normally feel daunted by eating a whole roast onion, this will cause you no trouble. The flavour is mild, fresh and sweet.**

1 Preheat the oven to 220°C/425°F/gas 7. Peel the onions and halve them lengthwise. Place in a single layer in a roasting tin. Brush generously with oil and sprinkle with the breadcrumbs, rosemary, sea salt flakes and pepper. Roast in the preheated oven for 30–35 minutes, until the onions are soft and the edges begin to blacken.

2 Using a swivel peeler, shave off thin flakes from the Parmesan. Sprinkle the flakes over the onions and serve at once.

## Sweet-and-Sour Onions with Dried Cranberries

**Serves 4–6 as a side dish**

*900g/2lb small pickling onions, unpeeled but roots trimmed*
*25 g/1 oz butter*
*1 tablespoon extra-virgin olive oil*
*1 bay leaf*
*4 cloves*
*5 tablespoons balsamic vinegar*
*1 teaspoon sugar*
*150–200 ml/5–7 fl oz chicken stock*
*sea salt and freshly ground black pepper*
*2 tablespoons coarsely chopped flat-leaved parsley or chives*
*3 tablespoons dried cranberries*

**This is a perfect accompaniment to roast pork, goose or turkey. The juices are deliciously rich and sticky, and taste almost like treacle. Cloves and cranberries cut the richness and prevent the dish from being overly cloying.**

1 Drop the onions into a large pan of boiling water. Bring back to the boil, blanch for 2 minutes, then drain. When cool enough to handle, remove the outer layer of skin. Leave the tips untrimmed as this will prevent the onions from disintegrating later on.

2 In a frying pan into which the onions can fit snugly in a single layer, heat the butter and oil with the bay leaf and cloves. When the butter is foaming, add the onions, stirring to coat, and simmer gently for a few minutes.

3 Add the vinegar and cook over moderate to high heat for 5 minutes, until reduced and sticky. Be careful not to let the pan dry out. Stir in the sugar, 150 ml/5 fl oz of the stock and some salt and pepper. Cook over low heat for 1 hour, stirring and turning the onions regularly, until the juices are reduced and syrupy and the onions are soft. Add more stock if the pan starts to look too dry.

4 When almost ready to serve, stir in the parsley and cranberries and simmer for another minute. Transfer the onions to a serving dish and season with plenty of pepper.

## ◂Leek & Green Peppercorn Tart

**Makes one 30 cm/12 inch square tart**

*375 g/13 oz rich shortcrust pastry, made with 225 g/8 oz flour and 140 g/5 oz butter, ¼ teaspoon each salt and sugar, 3 tablespoons iced water*
*115 g/4 oz Cheddar cheese, coarsely grated*
*115 g/4 oz mozzarella cheese, coarsely grated*
*1½ teaspoons dried green peppercorns, crushed*
*salt*
*3 small leeks, halved lengthwise and thinly sliced*
*55 g/2 oz pancetta, diced*
*2 teaspoons fresh thyme leaves*
*beaten egg yolk, to glaze*

**The smoky flavour of green peppercorns is delicious with leeks and cheese.**

**1** Combine the flour, salt and sugar. Work in half the butter until the mixture resembles coarse crumbs. Briefly work in the remaining butter, leaving it slightly unevenly mixed (this helps crisping). Lightly stir in the water to form a soft dough. Wrap in cling-film and chill for 1 hour.
**2** Preheat the oven to 240°C/475°F/gas 9. Thinly roll out the dough on a floured surface to form a 31 cm/12½ inch square. Carefully drape it over a rolling pin and place it on a baking sheet. Trim the edges neatly.
**3** Leaving a 2 cm/¾ inch border all round, scatter the cheeses evenly over the pastry. Sprinkle with the peppercorns and a little salt to taste, then add the leeks, pancetta and thyme.
**4** Fold the edges over the filling and brush with the egg yolk mixture. Bake for 20 minutes, until the pastry is golden and the leeks look slightly charred. Serve hot or warm.

## Three-Way Garlic Soup

**Serves 6**

*2 red onions, cut across into 1 cm/½ inch rings*
*4 large garlic cloves, unpeeled*
*3 red peppers, halved and deseeded*
*900 g/ 2lb ripe plum tomatoes, halved horizontally*
*2 tablespoons olive oil*
*1 teaspoon balsamic vinegar*
*425–550ml/15–18 fl oz chicken stock or vegetable stock, preferably home-made*
*salt and freshly ground black pepper*

**To serve:**
*garlic cream (page 126)*
*2 tablespoons chopped fresh coriander*
*garlic croûtons (page 126)*

**This soup of grilled garlic is served with garlic cream and croûtons, but it is not as overpowering as you would expect. The cloves are cooked whole, which mellows and sweetens the flavour.**

**1** Preheat a hot grill. Insert three wooden cocktail sticks horizontally into each of the onion slices to keep the rings in place. Place in a large grill pan along with the garlic cloves. Arrange the peppers, skin side up, and tomatoes in a second pan (or use one pan and grill in batches).
**2** Trickle the olive oil over the onions and tomatoes. Grill for 10–20 minutes, turning the onions once, until the garlic feels soft and the other vegetables are slightly blackened. Remove each vegetable as it becomes cooked. Remove the cocktail sticks from the onions.
**3** Squeeze the flesh from the garlic cloves and put in a food processor with the onions, peppers and tomatoes. Include any blackened bits of skin and scrape in the sticky juices from the grill pan. Process for about 2 minutes, then push the mixture through a sieve, pressing with the back of a spoon to extract all the liquid. Pour into a saucepan, add the vinegar and enough stock to achieve your preferred consistency. Season to taste, and simmer for 5 minutes.
**4** Pour the soup into bowls, swirl in some garlic cream and sprinkle with the coriander. Serve with garlic croûtons.

'Albina Vereduna'
'Boltardy'
'Burpee's Golden'
'Chioggia'
'Forono'

# BEETROOTS

If ever there was a vegetable whose fortune deserves turning around, then surely it must be the beetroot (*Beta vulgaris* subsp. *vulgaris*). Even with sugar levels as high as some sweetcorn varieties, a fresh earthy flavour equal to that of the best gourmet potatoes and rich, bright colours that rival those of peppers, the beetroot still fails to gain favour with the majority of cooks.

Colour is undoubtedly a problem – creamy soups and sauces turn an embarrassing shade of pink; fingers, chopping board and sink are inevitably stained with crimson; salads run the risk of unwelcome seepage. However, cooks need not limit themselves to red beetroot – there are golden beets, white beets and even some with stunning pink and white striped flesh, which don't have the seepage problem.

## A change of pigment

Unlike the carotene-induced reds and yellows of peppers, the colours in beetroot are caused by a group of pigments known as betalains. They come in red and yellow, and differences in their ratio determine a beetroot's colour. A deep red, for example, indicates low levels of the yellow pigment, while yellow shows that the red pigment is absent. Sometimes neither pigment is present, and the resulting colour is albino white.

## Cultivation and harvesting

The beetroot seed is actually a fruit, containing two or three seeds. Several seedlings may germinate from each fruit, forming a cluster of plants that can be thinned to provide a welcome treat of early season beet greens. There are also monogerm varieties, containing only one seed in each fruit but, unless the home gardener objects to thinning, there is little to gain by growing them.

With an efficient sowing strategy, beetroot can be harvested from early summer until late autumn. Early spring sowings may run to seed if the weather turns cold, though there are bolt-resistant varieties that can be grown at this time of year. Monthly sowings can be made until mid-summer, and the last roots either kept in the ground, covered with straw, or lifted and stored in a frost-free place.

## Buying and storing

It remains a mystery why beetroot are sold pre-cooked and, worse still, drowned in malt vinegar. Though not available all year round, fresh beetroot are far superior. Choose firm, smooth bulbs with the leaves attached. Separate the leaves before storing, leaving a short length attached to the bulb. Leave the whiskery roots intact. Unwashed bulbs will keep in a plastic bag in the fridge for up to a week, and the leaves, which are edible, for three days.

## Preparation

Wash the bulbs just before cooking, taking care not to damage the skin. Boil for 20 minutes to one hour until just tender. Drain, then nip off the stem and root, and peel away the skin.

Roasting concentrates the flavour beautifully. Loosely wrap untrimmed small-to-medium-sized bulbs in foil and roast at 190°C/375°F/gas 5 for 1–2 hours. When cool enough to handle, peel and trim. Alternatively, trim and peel them first, then roast without the foil, in preheated olive oil, perhaps with a sprig of thyme, a fat clove of garlic and a sliver of chilli. The bulbs become shrivelled, but they taste wonderful, as do the oily juices.

## Cooking

The cooking of Russia and Eastern Europe immediately springs to mind when considering beetroot. Borscht is the quintessential Russian dish; less well known is *chlodnik*, a chilled soup from Poland. Typical of these cuisines are ingredients which accentuate beetroot's slightly odd flavour – caraway, dill and juniper, buttermilk and sour cream. However, salty, pungent or bitter flavours are good partners too. Beetroot is delicious with smoked fish, herrings, horseradish and mustard, and it goes well with bitter leaves, such as rocket and chicory, giving rise to exciting ideas for salads.

Rather than treating beetroot juice as a problem, why not make the most of it? Team red beetroot with red ingredients. A stunning combination is thinly sliced blood oranges, beetroot and radicchio with a walnut oil dressing. Alternatively, try shredded red cabbage, beetroot, red onion and walnuts with a thick mustardy dressing.

Continuing with the red theme, a favourite dish is red flannel hash – hash browns made with grated beetroot, mashed potato, red peppers and onions. A beetroot risotto moistened with red wine is also good. We like a fiery casserole of beetroot, tomatoes, onions and kidney beans, seasoned with cayenne and turmeric, and topped with steamed beet leaves. We have even made beetroot kebabs; grilled over charcoal, they are surprisingly good.

## Nutrition

Beetroot is an excellent source of folate, one of the B vitamins, and contains a small amount of vitamin C.

## Varieties

For our tasting trials, the beetroots were cooked two ways: steamed and roasted wrapped in foil.

Two of our favourites are '**Moneta**' (not shown), a red monogerm variety, and '**Boltardy**', a bolt-resistant variety with a deep red colour. Both have a nice sweetness balanced against an obvious, though not overly earthy, flavour. They are delicious baked in foil with thyme. Another good variety is '**Chioggia**', which has a mild earthy flavour. The flesh is dramatically striped with red and white rings, which sadly fade to an almost uniform pink during cooking. Used raw, the colours can be shown off in a zinging salad – slice the flesh into very thin rounds and dress with a citrusy vinaigrette. The beautiful yellow '**Burpee's Golden**' has behind its sweet, earthy taste a slight off-flavour that leaves the mouth somewhat dry. The elongated deep red '**Forono**', though useful for slicing, is a bit bland overall. In contrast is '**Albina Vereduna**', the sweetest beetroot of the group, though the sugar tends to overwhelm its mildly earthy flavour.

# Baby Beetroots with Roasted Shallot & Chilli Dressing

**Serves 4**

*12 small yellow, white, pink and red beetroots*

*4 small handfuls of young dandelion leaves, chicory or rocket*

*sea salt flakes*

*coarsely ground black pepper*

**For the dressing:**

*2 banana shallots or 4 small ordinary shallots, unpeeled*

*2 New Mexican or Anaheim chillies*

*2 tablespoons lemon juice*

*8 tablespoons extra-virgin olive oil*

*salt and freshly ground black pepper*

**Deliciously sweet and mellow, baby beetroots contrast well with slightly bitter leaves and a mildly fiery dressing. The beets should be no more than 4 cm/1½ inches in diameter. You can, of course, use the dark red type if you don't have other colours.**

**1** Preheat the oven to 220°C/425°F/gas 7. Trim all but 1 cm/½ inch of stalk from the beetroots and any very long roots. Plunge into boiling water, bring back to the boil and simmer briskly for about 30 minutes, until the beetroots are just tender. Drain and leave to cool a little.

**2** While the beetroots are cooking, make the dressing: wrap each shallot in foil, place in a small roasting tin and roast in the preheated oven for 15 minutes. Add the chillies and roast for another 10–15 minutes, until the shallots feel soft and the chillies blacken and blister.

**3** Peel the shallots, roughly chop the flesh and put in a blender. Remove the skin and seeds from the chillies. Roughly chop the flesh of one of them and add this to the blender, together with the lemon juice, olive oil and seasoning. Purée until thick and smooth, and set aside. Cut the remaining chilli into very small neat squares.

**4** When the beetroots are cool enough to handle, slip off their skins, leaving the stalk in place and taking care not to damage the flesh. With a sharp knife, slice them in half lengthwise.

**5** Arrange the salad leaves on a serving plate with the beetroots on top. Sprinkle with the chopped chilli, sea salt flakes and a grinding of pepper. Spoon the dressing over and serve.

# SWEDES, TURNIPS & RADISHES

For some reason these humble vegetables have acquired a dubious reputation, and gardeners and cooks alike tend to approach them with a measure of disdain. Perhaps it is their intimate relationship with the soil which fuels such prejudices. After all, compare our attitude to these dirt-encrusted protuberances with the celebrity status we bestow on peppers and aubergines. Or, possibly, their strong sulphurous flavours are regarded as simply too crude for modern palates. Whatever the reason, they are due for some re-evaluation. They may have their limitations, but they are ready and waiting to come out of the cellar.

## Swedes

Swedes *(Brassica napus* Napobrassica Group*)* suffer from an image problem, particularly in northern Europe where their main use has been as animal fodder. At long last, however, food writers and chefs are beginning to use them more creatively, and varieties are being bred specifically for the table.

One of the biggest complaints concerns their bitter taste, produced by chemicals known as glucosinolates. Too high a concentration can result in an unpleasant flavour; too little, though, and the flavour can be bland. Like a good hoppy beer, the best varieties get the balance right.

Swedes also show considerable differences in texture, with some varieties having a notably harder flesh. From a culinary perspective, excessive hardness is an inconvenience, since the swede is more difficult to peel and chop, and will take longer to cook. From the gardener's point of view, however, hardness equals hardiness and enables the swede to better withstand freezing temperatures.

### Cultivation and harvesting

The best time to sow swedes is from late spring to early summer, since earlier sowings may bolt. Impatient gardeners can harvest their swedes as soon as the roots start to swell. However, postponing the harvest is likely to improve the flavour, since exposure to cooler temperatures tends to cause an increase in sugar levels.

### Buying and storing

Choose smaller specimens, if possible, as these are likely to be less fibrous. They should be free from cracks and feel weighty for their size. Swedes are amiable vegetables and can be stored for a month or more in a cool cellar or in the fridge without any obvious deterioration.

### Preparation

One of the pleasures of cooking with swedes is the tantalizing aroma that wafts up as you remove the skin. It's a clean smell, like a whiff of fresh horseradish. Peel your swede with a sturdy medium-sized sharp knife, but take care – the swede is tough and knives can slip. Cut it into quarters and then into wedges, slices or cubes depending on the recipe.

### Cooking

We find that steaming rather than boiling results in a better flavour and a less waterlogged texture, which is preferable if you want your pieces of chopped swede to keep their shape. However, boiling is fine for swede mash. Cut the swede into cubes, boil for 20–30 minutes, then mash to a coarse purée with butter or cream and plenty of salt and pepper. A seasoning of freshly grated nutmeg or root ginger does not go amiss.

Gentle roasting works, too, especially if the swede is lacking in flavour. The sugars caramelize and the flavour is intensified, resulting in golden nuggets that make a perfect accompaniment to a succulent roast.

Cut into thin strips and lightly blanched, swede is excellent stir-fried with an equally robust green leafy vegetable, such as kale or Savoy cabbage. It is also delicious raw in a winter coleslaw. Mix it with grated carrot, shredded green cabbage, plenty of chives and a mustardy dressing.

For a spectacular grilled swede salad, brush thin slices of uncooked swede with olive oil and cook on a ridged stove-top grill pan until tender and beginning to blacken. Serve it with peppery rocket, perhaps some smoked mackerel, and a creamy horseradish dressing. The combination of flavours is stunning.

'Melfort'
SWEDES
'Marian'
'Purple Top Milan'
TURNIPS
'Tokyo Cross'

## Nutrition

Surprisingly, swedes are a good source of vitamin C; but they are not particularly well endowed with other nutrients, although most are present in small amounts.

## Varieties

With few exceptions, swedes have creamy yellow flesh and a two-tone skin – purple on top and a dusky yellow where it sits below ground. One of our favourites, though, is '**Melfort**', a green-topped variety with good cold weather tolerance. It has a robust flavour, with noticeable sweetness and just a slight hint of bitterness. When steamed or boiled, the flesh remains firm and is not at all watery – perfect for mashing. When roasted, however, the flavour becomes quite intense and would probably be best appreciated only by committed swede enthusiasts.

Originally bred for animal fodder, '**Marian**' is nevertheless widely grown for culinary purposes. When the roots were steamed we found the flavour disappointing, with little sweetness and just a touch of bitterness. However, roasting brings out its qualities – try it in the tasty (and filling) Roasted Swede, Ginger and Black Bean Soup (page 141).

'**Ruta Otofte**' (not shown) has a high dry matter content, resulting in excessive denseness when raw and a firm, slightly fibrous texture when steamed or boiled. The flavour is improved by roasting. Given its winter hardiness, this is a worthwhile variety for growing in colder areas.

## Turnips

Turnips *(Brassica rapa* Rapifera Group*)* show considerably more refinement than the big, bruising swedes. Not only are the roots smaller, but they have soft, enticingly smooth, almost blemish-free skin. The delicately flavoured flesh is less dense than that of swedes, which makes the root sensitive to freezing temperatures.

Turnips are jet-setters, cropping up in the most unexpected cuisines. North African cooks use them to make colourful stews, fragrant with cumin and coriander. Together with their leafy tops, turnips are an essential ingredient in a classic meal-in-a-pot soup from Galicia in Spain. They are used all over the Middle East, stuffed, pickled, and in stews and soups. They also feature in Indian cuisine, in pickles and curries. The Japanese like them too, in stir-fries, pickles and soups, or plainly boiled and seasoned with sugar, vinegar and wasabi.

### Cultivation and harvesting

Turnips planted too early in spring may bolt if the weather turns cold. They give a quick return for your investment, with some varieties ready for pulling a little over a month after sowing. They are best harvested young, while still golf-ball-sized, though some of the varieties we tried were still delicious even when they had grown to 10 cm/4 inches in diameter.

### Buying and storing

Look for pale, firm, blemish-free specimens. The skin should be smooth rather than pitted and it should not feel spongy. Stored in the fridge or in a cool cellar or shed, fresh young turnips can be kept for a week and older ones for up to a month.

### Preparation

Smooth-skinned young turnips need nothing more than a quick rinse and their tops trimming. There's usually no need to peel small ones since the skin is paper-thin. Larger specimens may need peeling, but, as much of the flavour is concentrated in or near the skin, it's a pity to do so unless absolutely necessary. Boil them unpeeled and don't remove the skin unless if it is still tough or bitter after cooking.

### Cooking

Turnips can be interchanged with swedes in many recipes, but they need gentler cooking, particularly if young. Boil or steam them whole, halved or quartered, depending on size. Cook until just tender and serve tossed in melted butter, a generous showering of chopped parsley or chives, sea salt flakes and freshly ground black pepper. Alternatively, sizzle just-cooked turnips with butter, sugar and a little good stock, and bubble it down to a syrupy glaze (page 142).

Turnips are adept at mopping up other flavours and, in turn, they impart their own sweet mellow flavour to other ingredients. To experience this give-and-take quality, cook them in a moist braise alongside a richly flavoured duck or goose, or a succulent piece of lamb.

### Nutrition

Like swedes, turnips are well-endowed with vitamin C and contain modest levels of most other vitamins and minerals.

### Varieties

One of our favourites is the yellow-fleshed '**Golden Ball**' (not shown). Though some specimens can be unpleasantly bitter, this variety has, overall, a rich, almost buttery flavour, with a nice level of sweetness balanced with just the right amount of bitterness. The texture, too, is excellent: soft but not mushy. '**Purple Top Milan**' is a quick-growing flattened variety with white flesh. Although slightly hard and not especially sweet, it has a nice turnip flavour. Impatient gardeners wanting an attractive root might like to try the white-skinned '**Tokyo Cross**'. The flesh is fairly mild with some background butteriness. '**Green Globe**' (not shown) has a green-topped skin and white flesh that is moderately bitter and mildly sweet. It is soft-textured and slow-growing, and had the least to offer of the varieties we tried.

## Radishes

In the West, radishes *(Raphanus sativus)*, like swedes, have long tolerated an undeservedly inferior role. We tend to limit ourselves to the small-rooted types which gardeners use as gap-fillers and cooks only in salads. Fortunately, the radish enjoys more status in Asia. There the preference is for giant-sized varieties that are so exuberant they practically push themselves out of the ground. Some are the shape and size of a tennis ball, while others resemble giant carrots. They are also notable for their colours. As well as the familiar red-skinned, white-fleshed types, there are spectacular varieties with grass-green skin and crimson flesh, and those that are red or green all the way through. Oriental cooks take full advantage of their pungency and crisp texture, using them both raw and cooked in crunchy pickles, chutneys and salsas, stir-fries, curries and soups. All in all, large-rooted radishes provide a great opportunity for horticultural and culinary enjoyment.

### Cultivation and harvesting

Large-rooted radishes sown in the spring may bolt, although resistant varieties that do well at this time of year are available. In general, most varieties are better sown from summer to early autumn for harvesting in late autumn and winter.

The degree of pungency is strongly influenced by growing conditions and the age of the plant. For example, soils that have high clay or sulphur contents can push up the heat levels, and high temperatures may increase pungency to the point of causing sensory distress. Plants on the verge of bolting may also be hotter.

### Buying and storing

Look for large-rooted varieties in Middle Eastern and Asian food shops, good health-food stores or farmers' markets. Supermarkets are beginning to stock a wider range, including black Spanish radishes and the white Japanese 'daikon' or 'mooli' type.

Leaves should be removed before storing, since they encourage moisture loss and premature deterioration. Choose solid-looking unblemished specimens that do not feel spongy when pinched. Avoid excessively large specimens as these are often fibrous or pithy. Tightly wrapped in cling-film, large radishes will keep for several weeks in the fridge, although they will lose some pungency.

### Preparation

If using radishes unpeeled, remove the tops, root tip and any whiskery protuberances, and scrub them clean. Radishes are unpredictable and it is not until you take a bite that you can assess the heat scale. Pungency is concentrated in the skin and, often, at the growing tip, so peel and trim as necessary. Slice, dice or grate the radishes according to the recipe. If preparing ahead, keep the flesh moist as it quickly dries out.

### Cooking

Unless you are a radish addict, reserve the more fiery roots for cooked dishes and use the milder types raw.

Mild-tasting mooli or daikon can be finely sliced and added to salads, or grated and mixed with a little soy sauce to use as a relish. Grated and mixed with mayonnaise or crème fraîche, the more pungent, dense-fleshed winter radishes make a palate-tingling starter or accompaniment to cold meats and smoked fish. They also make good pickles.

Sliced wafer-thin, radishes are delicious floated in a clear Japanese-style noodle soup. Alternatively, cut them into cubes and add these to curries and stir-fries along with other vegetables. The heat from cooking seems to tone down excessive pungency.

For a delicious mint-flavoured stir-fry, toss diced radish in hot oil with mustard seeds, chopped fresh chilli, sea salt flakes and plenty of shredded mint.

### Nutrition

Being over 90 per cent water, radishes contain relatively low levels of most nutrients. Eaten in sufficient quantities, however, they provide valuable amounts of folate and are a good source of vitamin C.

### Varieties

We sowed a number of varieties of large-rooted radishes in late summer and early autumn, both outdoors and in polytunnels. They were harvested from late autumn to mid-winter. In common with their smaller relatives, these large varieties have a characteristically pungent flavour. We tasted the skin and flesh separately and, without exception, found the skin to be more pungent. Our comments on flavour and texture, therefore, are usually limited to the flesh alone.

Highly recommended is the roundish red-skinned '**Cherokee**' (not shown). Its ice-white flesh has an excellent dense but crisp texture, a modicum of pungency and a nice touch of sweetness. Another favourite is '**Minowase Summer Cross**', a quick-growing white mooli type. The flesh is crisp and juicy, sweet and slightly bitter but not pungent. We like it grated in a coleslaw.

Some varieties are worth growing not only for their flavour but also for the magnificent colours they bring to a dish. '**Misato Green**' is a real stunner. It is elongated, with skin and flesh that are green on the upper part and white on the lower. The flesh is crisp and juicy, slightly pungent and sweet. Its partner, '**Misato Rose**', has a light red skin and flesh with a good flavour, some sweetness and a touch of pungency. The round '**Mantanghong**' (not shown) has a green skin and an exquisitely dark pink crisp flesh with a slightly sweet flavour. Together, these three varieties make a spectacular salad (page 143); or, for a brilliant salsa, cut them into small dice and mix them with chopped spring onion, cumin seeds, chopped fresh coriander, lime juice and sea salt flakes.

We have been less impressed by '**China Rose**' and '**Violet de Gournay**'. Though some roots are sweet, 'China Rose' has too mild a flavour for our liking; 'Violet de Gournay' is rather unpleasantly flavoured, with bark-like, excessively pungent skin. '**Black Spanish Round**' suffers from a similar skin complaint, but once peeled reveals a dense, sweet, tolerably pungent and slightly bitter flesh that is good in pickles and cooked dishes. It stores well and the pungency tones down after a few weeks in the fridge.

## Roasted Swede, Ginger & Black Bean Soup

**Serves 4–6**

*2 teaspoons coriander seeds*
*2 teaspoons cumin seeds*
*2 teaspoons dried oregano*
*550 g/1¼ lb swede, peeled, quartered and cut into wedges*
*1 large sweet onion, cut into thick wedges*
*3 tablespoons olive oil*
*1 teaspoon ground ginger*
*3 fat garlic cloves, unpeeled*
*225 g/8 oz black beans, soaked overnight*
*1 litre/1¾ pints chicken stock, preferably home-made*
*salt and freshly ground black pepper*
*6 tablespoons chopped flat-leaved parsley*
*6 tablespoons crème fraîche or soured cream*

**Roasting is a good way of bringing out the flavour of mild-tasting swedes. Garlic emphasizes their sweetness, and the beans add a mellow, earthy touch. This soup will have you staggering from the table.**

**1** Preheat the oven to 190°C/375°/gas 5. Dry-fry the coriander and cumin seeds in a small pan over a moderate to high heat until fragrant. Add the oregano and fry for a few seconds more. Quickly remove from the heat and grind to a coarse powder using a pestle and mortar.

**2** Put the swede and onion wedges in a large bowl and toss with the oil, the ginger and a tablespoon of the toasted spices until evenly mixed. Place the vegetable wedges and the garlic cloves in a single layer in a large roasting pan, spreading them out well. Roast for 15 minutes, then remove the garlic and leave the swede and onion to roast for another 25 minutes, turning occasionally, until they are soft and slightly blackened.

**3** Meanwhile, drain the beans and put in a pan with fresh water to cover. Bring to the boil and boil rapidly for 20 minutes, then continue to cook at a brisk simmer until tender – about a further 20 minutes. Drain and set aside.

**4** Peel the garlic and put in a food processor with the roasted swede and onion wedges. Add 600 ml/1 pint of the stock and process to a chunky purée.

**5** Scrape the purée into a saucepan. Add the beans and the remaining stock and spice mixture. Season generously with salt and pepper. Bring to the boil, then simmer, covered, for 20–30 minutes. Swirl in the parsley and cream just before serving.

# Glazed Turnips with Coriander & Orange

**Serves 4 as a side dish**

*900 g/2 lb turnips, left whole and unpeeled if small, otherwise peeled and quartered*
*40 g/1½ oz butter*
*1 teaspoon coriander seeds, crushed*
*2 tablespoons sugar*
*finely grated zest of 1 small orange*
*150 ml/¼ pint chicken or vegetable stock, preferably home-made*
*sea salt flakes*
*coarsely ground black pepper*
*1 tablespoon orange juice*
*2 tablespoons chopped fresh coriander or flat-leaved parsley*

**The sweet syrupy juices in this dish contrast well with the slightly bitter turnips, and the orange juice pulls it all together. This is delicious as an accompaniment to a roast, or served as a vegetarian main course with a dish of steamed leafy greens or a bitter leaf salad.**

**1** Plunge the turnips into a large pan of boiling salted water. Cook for 3–4 minutes until barely tender, then drain.

**2** Melt the butter in a frying pan into which the turnips will fit snugly in a single layer. As the butter is melting, sprinkle in the coriander seeds. When the butter is foaming, slip in the turnips, sprinkle with the sugar and orange zest, and cook over moderate to high heat, turning, until the turnips are beginning to colour at the edges.

**3** Pour in the stock, season generously, then simmer briskly for a few minutes until the liquid is reduced and syrupy. Stir in the orange juice and sprinkle with the fresh coriander or parsley.

# Rainbow Radish Salad with Carrots & Kohlrabi

**Serves 6 as a starter**

*85 g/3 oz mooli*
*1 large red-fleshed radish, such as 'Mantanghong' or 'Misato Rose'*
*1 large green-fleshed radish, such as 'Misato Green'*
*6 small radishes*
*2 carrots*
*1 kohlrabi*
*sea salt flakes*
*8 tablespoons extra-virgin sunflower oil*
*1 teaspoon black sesame seeds or onion seeds*
*a few drops of toasted sesame oil*

**If you don't have green- or red-fleshed large radishes, you can, of course, make up the difference with small red-skinned radishes and extra mooli.**

1 Scrub the radishes clean and peel the carrots and kohlrabi. Using a mandolin or very sharp knife, slice all the vegetables into paper-thin rounds. Arrange them on individual plates or in a shallow serving dish. Sprinkle lightly with sea salt flakes.

2 Put the sunflower oil and the sesame seeds in a small saucepan and heat until the seeds begin to pop. Pour the hot mixture over the radishes, sprinkle with the sesame oil and serve while the dressing is still warm.

# CELERIAC & KOHLRABI

If an ugly vegetable contest were ever to be held, celeriac *(Apium graveolens* var. *rapaceum),* would be the hands down winner. With a sombre green and brown pock-marked skin, it has a rough-and-ready look. In contrast, kohlrabi *(Brassica oleracea* Gongylodes Group) presents a picture of purity, with smooth pale green or purple skin powdered with a light white blush, broken only by elegant upward-growing leaves.

## Cultivation and harvesting

Celeriac is a cool-temperature crop sown in the spring and harvested in the autumn when the roots are about 8–13 cm/3–5 inches in diameter. In mild areas, they can be left in the ground until the following spring.

Like celeriac, kohlrabi is a cool-season crop. However, it takes less time to mature, and three or four successive crops can be grown for continuous harvests from late spring to mid-winter. For an early harvest, it might be worth growing the first crop in a polytunnel or greenhouse.

Kohlrabi becomes tough and fibrous as it matures, though modern breeding has created less fibrous varieties. We harvest it when the stems have swollen to around 5 cm/2 inches in diameter.

## Buying and storing

Celeriac should feel firm and heavy for its size. To make preparation easier, go for the smoothest and cleanest. Choose small kohlrabi – no bigger than a tennis ball – and reject any that are badly scarred or discoloured.

Celeriac can be stored for a week or more, tightly wrapped in cling-film and kept in a cool place. Kohlrabi will keep for up to a week in the fridge in a plastic bag.

## Preparation

Scrub celeriac under running water to remove excess dirt. With a very sharp knife, cut it into quarters then trim away the skin and root. Once cut, the flesh discolours very easily so plunge it into a bowl of water acidulated with lemon juice, or, if you are using it raw in a salad, stir it straight into the dressing.

Kohlrabi needs peeling unless it is exceptionally young and tender. Older specimens may have a fibrous layer immediately below the skin and this must also be removed.

## Cooking

Celeriac must be one of the most underrated of all the winter vegetables. Since the flesh is not only extremely tasty but versatile too, it deserves a better deal.

Finely grated, the raw flesh is delicious added to leafy salads, or mixed with a mustardy mayonnaise to make celeriac *rémoulade.* It's also good with horseradish cream (page 162). Cooked and puréed, it is a fine accompaniment to a roast saddle of wild boar or venison. It makes a good soup: top with fresh dill and a swirl of soured cream. A few chunks in a hearty winter casserole add a refreshing flavour. Combine it with sweet-tasting roots such as squash or sweet potatoes and serve with lamb knuckles (page 146). Or try roasting pieces of celeriac around a leg of pork or with duck.

Kohlrabi needs gentler treatment if its delicate flavour is not to be swamped by other ingredients. We like a simple salad of diced kohlrabi tossed with grated lemon or lime zest, lemon juice, olive oil and some chopped fresh lovage. Or serve separate mounds of finely grated kohlrabi, celeriac and white or green radishes, with a walnut oil dressing made with 1 tablespoon of lemon juice, ½ teaspoon each of sugar and Dijon mustard, a pinch each of celery salt, sea salt and freshly ground black pepper, 3 tablespoons of light olive oil and 1 tablespoon of walnut oil.

Cooked kohlrabi is best when either steamed or boiled until just tender; serve simply with melted butter and lots of parsley or dill for a bit of colour. Alternatively, it can be cut into batons and lightly stir-fried with shredded greens and a dash of lemon juice.

## Nutrition

Celeriac and kohlrabi are useful sources of carbohydrate and dietary fibre, particularly the soluble type which is believed to help lower blood cholesterol levels. They contain important minerals, including potassium, calcium, magnesium and iron, and are surprisingly generous providers of folate and vitamin C.

## Varieties

Celeriac is not a vegetable that we grow; instead we buy it from market gardening friends who grow '**Brilliant**' and '**Balder**' (not shown). Both are reliable croppers with a good celery flavour.

Kohlrabi, however, is becoming one of our favourite vegetables. We taste-tested a number of tunnel-grown varieties harvested in the late spring. Both raw and cooked, the green-skinned '**Eder**' (not shown) was a revelation – full-flavoured, sweet and 'minty', with a fat, juicy texture. Try it in a salad with lemony sorrel and crisp apples (page 146). Though not quite as well-flavoured or sweet as 'Eder', '**Purple Danube**' (not shown) also gets high marks. Tasted raw, green-skinned '**Rowel**' (not shown) had a crisp, apple-like texture. Cooked, it had a sweet, well-rounded flavour. '**Purple-skinned Kolibri**' (not shown) was mild-flavoured but fairly sweet. '**Purple Vienna**' and '**Green Vienna**', both older varieties, were mild-flavoured and not especially sweet.

Celeriac 'Brilliant'

'Green Vienna'

'Purple Vienna'

KOHLRABI

## Lamb Knuckles with Braised Celeriac & Sweet Potatoes ▸

**Serves 6**

*1 teaspoon cumin seeds*
*2 teaspoons coriander seeds*
*1 tablespoon sesame seeds*
*2 teaspoons dried oregano*
*freshly ground black pepper*
*6 lamb knuckles weighing about 225 g/ 8 oz each*
*2 tablespoons groundnut oil*
*2 onions, roughly chopped*
*6 large garlic cloves: 3 finely chopped, 3 peeled and halved*
*400 g/14 oz can of chopped tomatoes*
*salt*
*850 ml/1½ pints meat stock, preferably home-made*
*70 g/2½ oz butter*
*1 tablespoon finely chopped fresh rosemary*
*2 potatoes, cut into chunks*
*1 large celeriac, cut into chunks*
*2 orange-fleshed sweet potatoes, thickly sliced*
*rosemary sprigs, to garnish*

**Ground toasted seeds thicken the juices and add a spicy flavour. Serve with buttered kale or Savoy cabbage.**

**1** Put the seeds in a small frying pan. Dry-fry until they start to splutter, sprinkle in the oregano and fry for a few seconds more. Remove from the heat and crush with a pestle and mortar.

**2** Generously season the lamb with pepper. Heat 1 tablespoon of the oil in a large heavy-based casserole, and brown the lamb evenly. Remove with a perforated spoon.

**3** Add the onions to the pan and gently fry for about 5 minutes. Add the chopped garlic and the seed mixture and fry for another minute. Stir in the tomatoes and season to taste.

**4** Return the meat to the pan and pour in 600 ml/1 pint of the stock. Bring to the boil, cover tightly and cook in a preheated oven at 180°C/350°F/gas 4 for 1¾–2 hours, until very tender.

**5** After the lamb has been cooking for a little over an hour, heat the remaining oil and half the butter in a large pan over a medium heat. Add the rosemary and the root vegetables in small batches and cook gently until just beginning to brown at the edges. Season each batch as you cook them. Don't overcrowd the vegetables, otherwise they will steam instead of browning. Add more butter as necessary and turn the vegetables so that they colour evenly.

**6** When all the vegetables are browned, reduce the heat and add the halved garlic. Cover and cook for about 3 minutes, until lightly coloured.

**7** Return all the vegetables to the pan and pour over the remaining stock. Cover and cook gently for 10–15 minutes, until the vegetables are tender but still holding their shape. Uncover the pan, raise the heat a little and cook rapidly to reduce the juices, turning the vegetables carefully.

**8** Using a perforated spoon, transfer the meat to a warmed serving dish and surround with the vegetables. Garnish with rosemary sprigs. Strain the meat juices into a small saucepan, blotting up any surface fat with paper towels. Reheat or, if they need thickening, bubble down over high heat. Check the seasoning and pour the juices over the lamb.

## Kohlrabi, Apple & Walnut Salad

**Serves 4 as a starter**

*3 small kohlrabi, peeled, sliced and cut into bite-sized segments*
*2 crisp red-skinned apples, diced*
*lemon juice*
*handful of young sorrel, torn into shreds*
*25 g/1 oz walnut halves, broken into pieces*
*walnut oil dressing (page 144)*

**Kohlrabi's fresh flavour goes well with the lemony sharpness of sorrel. If sorrel is unavailable, use watercress or rocket instead.**

**1** Combine the kohlrabi and apples in a serving bowl. Sprinkle with a little lemon juice to prevent browning. Add the sorrel and walnuts.

**2** Whisk together the dressing ingredients. Just before serving, pour the dressing over the salad and toss well.

# FENNEL

Fennel is the fashion queen of vegetables, elegantly draped in willowy frond-like leaves and perfumed with the warm, penetrating fragrance of anise. Poised gracefully in the garden, fennel offers unspoken promises of eating pleasure, and brings a sense of style to the table.

For all its elegant demeanour, fennel has something of a split personality, presenting the gardener and cook with two distinctly different forms. The well known Florence fennel *(Foeniculum vulgare* var. *azoricum*) has big, bold bulbs closely hugging the ground as if to provide ballast for the tall feathery leaves.

Another type of fennel *(F. vulgare)* is a thin waif of a plant that cannot seem to muster the energy necessary for bulb formation. Appearances are deceptive, however, and this is actually a vigorous and productive herb that continuously sends up new leaves throughout the growing season. These can be harvested periodically and used in the same way as any other fresh herb.

## Cultivation and harvesting

The non-bulbing fennels are perennials that faithfully emerge year after year despite minimal care. There are green and bronze varieties, both of which can be started from seed or by clump division. We like the bronze type, as much for its ornamental value as for its culinary qualities.

The bulbing type is also a perennial, though it is not managed as such. Instead, harvesting takes place when the bulbs are big and fat, usually in the summer and autumn in the case of outdoor plants. Planting in the spring can cause bolting before the bulbs reach full size. Unless bolt-resistant varieties are grown, you should delay sowing until temperatures warm up, or try planting the earliest crops in tunnels.

When harvesting bulbing fennel, try leaving some of the base attached to the roots. Left to their own devices, the plants will sprout new shoots, which then develop into baby fennel – a delicacy much loved by cooks.

## Buying and storing

Choose smooth, tightly packed bulbs, preferably with a few fresh green fronds attached. Fresh bulbs should be a luminous white with a faint sheen. Reject bulbs that are beginning to brown or look as if they have been excessively trimmed, as they will undoubtedly be past their best.

Fennel keeps well and will last for up to a week in the salad drawer of the fridge.

## Preparation

Trim the base and cut off the hard round stalks, saving any fronds attached. Chop these finely and use instead of the fennel herb to give a subtle aniseed flavour to salads, soups and fish dishes. Discard any tough outer layers and slice according to the recipe. For cooked dishes, it is usually better to halve or quarter the bulb lengthwise and cut out the core.

## Cooking

With the exception of tomatoes, no vegetable could be more evocative of Italian cuisine than fennel. The bulb is served raw or cooked – grilled, braised, baked and steamed – and in combination with numerous ingredients. Depending on the region, the leaves are boiled in water which is then used to cook pasta; and slivers of the bulb may be served after a meal as a digestive, either alone or with other raw vegetables.

Fennel is perhaps best appreciated raw. It makes a punchy salad ingredient – a few slivers added to lettuce add welcome texture and flavour. Even better is a salad of thinly sliced bulbs and a heavy shower of chives, tossed in a thick lemony vinaigrette. Alternatively, try fennel with slivers of red onion, or orange segments – or both. Watercress and rocket make good partners too, as do pears, walnuts and crumbly white cheeses. Crisp chunks of fennel are delicious as a crudité, dipped in good home-made mayonnaise or served with *bagna cauda* – a warm bath of olive oil, garlic and anchovies.

When cooked, the flavour becomes more subtle but the texture still remains pleasantly crunchy. We like the bulbs roasted in a hot oven until slightly blackened at the edges, then sprinkled with plenty of lemon juice and flat-leaved parsley (page 150).

The liquorice-like flavour perfectly offsets rich and oily fish such as sardines, mackerel and red mullet. Scatter crushed fennel seeds over the fish and stuff the cavity with leaves. A dash of Pernod will accentuate the flavour even more. Or try a gutsy Sicilian pasta sauce of sardines and anchovies (page 150) in which bulb, seeds and leaves are all used to good effect.

A particularly pleasing combination is fennel and radicchio, or Treviso chicory; the bitterness of the chicory is excellent with the sweetish flavour of the fennel. Slice a large head of each and fry them gently in olive oil with chopped onion until tender – delicious with pasta.

## Nutrition

Fennel is a useful source of folate, and contains some cellulose – a form of dietary fibre. It provides small amounts of potassium and zinc, as well as carotene, niacin (vitamin $B_3$) and vitamin C. Fennel is also valued for its digestive properties, which are attributed to anethole, one of the essential oils responsible for its flavour.

## Varieties

We have been loyally growing the bulbing '**Zefa Fino**' for almost ten years. It is supposed to be bolt-resistant, though we have never planted it early enough in the season to verify this. The bulbs can sometimes be a little flat in shape, and we therefore keep promising ourselves to try other varieties, such as '**Cantino**' (not shown) – also supposedly bolt-resistant – or '**Rudy**' (not shown), that are said to produce large, portly bulbs. Old habits die hard, however, and we have hitherto been reluctant to abandon an old friend.

## Red Mullet with Roasted Fennel▸

**Serves 4**

*3 large fennel bulbs with leaves and stalks*
*4 medium-sized red mullet, cleaned and gutted*
*olive oil for brushing*
*sea salt flakes*
*freshly ground black pepper*

**To garnish:**

*a few robust salad leaves such as mizuna or lamb's lettuce*
*lemon wedges*

**Fennel and red mullet make a good team – their flavours complement each other perfectly. Small sea bass would be another option.**

**1** Preheat the oven to 220°C/425°F/gas 7. Chop the fennel leaves finely and set aside 2 tablespoons. Chop the stalks and remaining leaves, mix together and reserve.
**2** Quarter the fennel bulbs and remove the core. Brush the bulbs generously with olive oil, and sprinkle with sea salt and plenty of pepper. Place them in a single layer in a roasting tin and roast for 20–30 minutes, until they are tender and beginning to blacken at the edges.
**3** While the fennel is cooking, score the fish on each side with a sharp knife and place it in a grill pan. Stuff the cavities with the chopped fennel mixture. Sprinkle with olive oil and season with salt and pepper. Place under a very hot grill for 4–5 minutes each side, then grill for another 4–5 minutes at a gentler heat (or farther away from the heat source) until the flesh is just opaque.
**4** Transfer to a warm serving dish and sprinkle with the reserved fennel leaves. Garnish with salad greens and lemon wedges, and serve with the roasted fennel.

## Bucatini with Fennel & Fresh Sardines

**Serves 4**

*½ teaspoon saffron threads*
*1½ tablespoons tomato purée*
*25 g/1 oz raisins*
*8 tablespoons olive oil*
*½ small onion, finely chopped*
*1 fennel bulb, finely chopped*
*1 large garlic clove, crushed*
*½ teaspoon fennel seeds, crushed*
*6 anchovy fillets, chopped*
*450 g/1 lb fresh sardines, filleted*
*40 g/1½ oz pine nuts, toasted*
*freshly ground black pepper*
*salt*
*350 g/12 oz bucatini*
*4 tablespoons toasted breadcrumbs*

**This is an adaptation of a classic Sicilian pasta dish. If you grow fennel as a herb, boil up a large bunch of leaves and stalks – or use a mixture of leaves, stalks and chopped trimmings from bulbing fennel – and use the water to cook the pasta. Bucatini is a long, round, hollow pasta, just right for a robust sauce such as this. You could use a wide ribbon type instead.**

**1** Make a solution of saffron and tomato purée with 225 ml/8 fl oz of hot water. Cover the raisins with more hot water and soak for 15 minutes, then drain and chop roughly.
**2** Heat the oil in a large frying pan, and gently fry the onion and fennel until just soft. Add the garlic, fennel seeds and anchovies, and fry for 1–2 minutes, mashing the anchovies to a paste. Add the sardines and fry briefly on each side.
**3** Stir in the pine nuts, raisins and saffron solution. Season with pepper but go easy on the salt. Simmer for 5–7 minutes until the liquid reduces and the sardines start to disintegrate a little.
**4** Meanwhile, cook the pasta in boiling salted water, or fennel water, until it is *al dente*. Drain and transfer to a warmed serving dish. Toss with the sauce and sprinkle with the breadcrumbs and fennel leaves.

# CARROTS

A good example of how perception of flavour can be strongly influenced by the other senses, such as sight and smell, was provided by a group of blindfolded eight-year-olds at our local school. The children were offered both white and orange carrots *(Daucus carota)* and, almost to a child, agreed that both types were equally delicious. Reactions changed, though, when the test was repeated without blindfolds. Many of the children were unable to eat the white carrots, claiming that they tasted 'horrible'.

Colour notwithstanding, sugar content wields a strong influence on carrot flavour. Also playing a pivotal role are volatile terpenes – chemicals responsible for the typical carrot flavour. Breeders make great efforts to strike the right balance of terpene and sugar levels. If there is too much terpene, the flavour can be unpleasantly harsh. Too far the other way, though, and the roots are hopelessly bland.

## Cultivation and harvesting

Carrots come in various shapes and sizes, and must be carefully chosen to match the garden soil. For example, long varieties might become stunted in shallow soil, but short round varieties do well. Carrots are also fussy when it comes to climate, and out-of-season sowings of the wrong variety may prematurely bolt.

Carrots can be harvested in late spring from sowings made in polytunnels in the previous autumn and again in mid-winter. Successive sowings can also be made outdoors from spring to early summer, providing much-appreciated roots from late summer and into the winter.

## Buying and storing

Look for crisp smooth carrots with bright orange skin. Steer clear of limp specimens and those with bruises or cracks. Though leaves may give an impression of freshness, water from the roots is actually lost through them, so remove the tops before storing. Loosely wrapped in a plastic bag, carrots will keep for at least a week in the fridge. Baby carrots should be used within a day or two.

## Preparation

Although most of the nutrients and flavour lie just below the skin, it is usually necessary to peel carrots unless they are very young. Use a swivel peeler and shave away a paper-thin amount of skin. Large old carrots may have a tough woody core which no

'Nairobi'
'Minicor'
White

amount of cooking will soften. If you suspect this is the case, quarter the carrot lengthwise and prise out the core. Carrots can be cut into any shape that takes your fancy. Remember the larger the pieces, the longer the cooking time.

## Cooking

In terms of flavour, one of the most rewarding methods is to fry thin bias-cut slices in a couple of tablespoons of vegetable oil over very high heat. Keep the slices in a single layer and turn them regularly with tongs. The result is a concentration of flavour, a caramelization of sugars and, with the slight blackening and crisping that takes place, a visually attractive finished dish. Roasting carrots in the oven, either around a joint of meat or mixed in with potatoes, achieves a similar concentration of flavour.

Braising is another method that brings out the natural sweetness of carrots. Cook small whole carrots, or sliced larger ones, in a small amount of water with some butter, a pinch of sugar, and salt and pepper. The water will eventually evaporate, leaving a rich syrupy glaze. Sharpen the flavour with a few drops of lemon juice, and, if you wish, sprinkle with chopped mint or chives.

Carrots go happily with anything that either blends or contrasts with their natural sweetness. For example, shaved raw carrots are wonderful in a salad with sweet fresh coconut, mangoes, fresh coriander and a lime juice dressing (page 154); and grated carrot cooked in coconut milk with dried fruits is a wickedly rich dessert. For a sweet-sour flavour, try a stir-fry of carrots, perhaps with some burdock root (page 168), seasoned with a few drops of vinegar or Japanese *umeboshi* seasoning – a salt-pickled plum juice widely available from good healthfood shops or the larger supermarkets. A sweet and pungent relish of grated carrot, fresh horseradish and soured cream is also good.

'Early French Frame'

## Nutrition

As their name suggests, carrots are a major source of carotenes, including not just the better-known beta-carotene, but alpha-carotene and lutein too, valued for their cancer-fighting properties. Eating a single carrot a day is thought to reduce the risk of lung cancer by 50 per cent. A daily carrot is also believed to improve night vision.

Like all root vegetables, carrots contain energy-boosting carbohydrates, as well as dietary fibre. Young carrots contain a small amount of potassium and magnesium, but the levels in older ones are insignificant. Carrots also contain small but useful amounts of vitamins C and E – both cancer-fighting antioxidants – and vitamin $B_6$.

## Varieties

Despite the importance we attach to flavour, the growing environment is our main concern when choosing varieties.

We have had very good results from autumn and winter tunnel sowings of the bolt-resistant '**Primo**' (not shown) and '**Nantucket**' (not shown). For outdoor sowing at other times of year, we grow '**Comet**' (not shown), which produces a good crop of wedge-shaped roots on our clay soils. Another vigorous grower is the **white carrot**, used as animal fodder in France. It has a surprisingly good flavour, but lack of carotene reduces its health benefits. To compensate, combine it with orange carrots in Two-Carrot Soup with Lovage (page 154), swirling the two colours together, or mix very thin slices of each in a salad with toasted pine nuts and Witloof chicory.

The short, almost round '**Early French Frame**' and '**Parmex**' (not shown) are useful for heavy or shallow soils, though their yields are low. The bolder roots of '**Nairobi**' are best grown in sandy soils where they can stretch out to their full size without restriction. Carrots also grow well in tubs, and we especially like the small baby roots of '**Minicor**'.

## Two-Carrot Soup with Lovage▶

**Serves 4**

*55 g/2 oz unsalted butter*
*4 tablespoons chopped lovage*
*1 white onion, chopped*
*2 potatoes, sliced*
*225 g/8 oz orange carrots, sliced*
*225 g/8 oz white carrots, sliced*
*850 ml/1½ pints vegetable or chicken stock, preferably home-made*
*salt and freshly ground black pepper*
*small lovage leaves, to garnish*

**White carrots have an intense flavour which goes well with lovage. If you don't have any lovage, use tarragon or chervil instead. You can, of course, make a single-coloured soup using 450 g/1 lb of orange carrots.**

**1** Divide the butter and lovage between two saucepans and melt the butter over medium-low heat. Add an equal amount of onion and potato to each pan. Put all the orange carrots in one pan and all the white carrots in the other. Cover and sweat for 10 minutes, stirring occasionally.
**2** Divide the stock between the two pans and season with salt and pepper. Bring to the boil, then simmer for 15 minutes.
**3** Liquidize the mixtures separately and return to their pans. Reheat gently and check the seasoning.
**4** Slowly pour some of each coloured soup into warmed bowls. Swirl the colours and garnish with a lovage leaf.

## Carrot, Coconut & Mango Salad

**Serves 4 as a starter or side dish**

*4 carrots, cut into 5 cm/2 inch lengths*
*75 g/2¾ oz fresh coconut, thinly sliced, or 40 g/1½ oz toasted coconut flakes*
*1 small mango, diced*
*2 tablespoons chopped fresh coriander*
*finely grated zest of 1 lime*

**For the dressing:**

*2 tablespoons lime juice*
*1 teaspoon muscovado sugar*
*½ small chilli, deseeded and very finely chopped*
*½ teaspoon cumin seeds, crushed*
*½ teaspoon salt*
*freshly ground black pepper*
*3 tablespoons extra-virgin sunflower oil or light olive oil*

**Having a similar texture and sweetness, carrot and coconut are a good combination. Mango, lime and coriander add a touch of sharpness. If you don't have a fresh coconut, use dried flaked coconut instead – easily found in health food stores but not always in supermarkets. Desiccated coconut will not do.**

**1** Using a swivel peeler and working from opposite sides, or from four sides if it is a fat carrot, shave the carrot pieces into wide ribbons, discarding the woody core. Combine the carrots, coconut, mango, coriander and lime zest in a bowl.
**2** Whisk the dressing ingredients and pour over the carrot mixture, tossing well. Leave to stand at room temperature for at least 30 minutes before serving, to allow the flavours to develop.

# JERUSALEM & CHINESE ARTICHOKES

Let's address the issue right from the beginning: Jerusalem artichokes cause flatulence. They have the misfortune of containing inulin, an indigestible carbohydrate, which passes innocuously through the upper regions of the digestive tract until it reaches the colon or large intestine. Here, things change: resident bacteria start feeding on inulin, resulting in an insidious build-up of carbon dioxide, methane and hydrogen, which in turn creates pressure in the bowel. The gases have only two escape routes, and they take the nearer one.

Despite this unwelcome characteristic, Jerusalem artichokes *(Helianthus tuberosus)* are an excellent winter vegetable to be enjoyed on an occasional basis. They have a distinctive earthy flavour similar to that of globe artichokes, which comes as no surprise since both are members of the Asteraceae family. However, the Chinese artichoke (*Stachys affinis),* sometimes referred to by the French name, *crosnes*, comes from yet another family – the Lamiaceae – which includes mints and basils. Even though of different ancestry, the Chinese artichoke still causes flatulence, in this case arising from stachyose, the same carbohydrate as that found in dried beans.

## Appearance

The Jerusalem artichoke is something of a plain Jane in the vegetable world. The skin is an undistinguished beige – though there are more colourful varieties with earthy pink or reddish skin – while the flesh is off-white, sometimes verging on grey. The Chinese artichoke is slightly more glamorous, resembling a tiny necklace of barrel-shaped ivory rings, each separated by a groove.

## Cultivation and harvesting

Both Jerusalem and Chinese artichokes are propagated simply by pushing a few tubers into well-prepared soil in the early spring and digging up their much-expanded numbers in the autumn. Remember, however, that tuberous artichokes are like a bad reputation: easy to get, hard to lose. Harvesting must be meticulous, since any that are missed will regrow the following year.

In mild areas Jerusalem artichokes are happy to be left in the ground throughout the winter. Should you choose to dig up your crop all at once, unwashed tubers will keep in a plastic bag in the fridge for a surprisingly long time. We have stored them for as long as three months, with little deterioration in eating quality.

Chinese artichokes, however, do not store well and are better left in the ground and harvested as required. They should be kept moist after digging up, since dried soil can be difficult to remove from their narrow grooves. Cleaning is simply a matter of rinsing the tubers under cold running water, patting them dry and laying them on a cloth with some sea salt. A gentle rub removes the soil, except from the deepest grooves.

## Buying and storing

Choose the largest and smoothest Jerusalem artichokes, since these will be easiest to peel. Avoid specimens that are soft and flabby as they are likely to be old. Also to be avoided are those that are bruised or broken. Chinese artichokes should look fresh and clean with no discoloration or dried tips. Store them in the fridge, inside a polythene box lined with damp paper towel, and use as soon as possible, preferably within a day or two.

Depending on freshness, Jerusalem artichokes bought from shops will keep in a plastic bag in the fridge for two to three weeks, but use them at once if you see signs of sprouting. You may find that their pleasant background sweetness increases when stored at low temperatures (see Potatoes, page 171, for 'reconditioning').

## Preparation

Scrub Jerusalem artichokes using plenty of cold water and slice off any whiskers and small knobs. If they have had a preliminary clean (see above) or are shop-bought, Chinese artichokes simply need a quick rinse. If the tips look a bit dry, nip them off. Peeling is unnecessary and impossible.

The conventional advice concerning Jerusalem artichokes is to immerse the tubers in acidulated water as you peel them, to prevent them from browning. In our experience browning isn't a particular problem as long as you cook them straight away – so why waste a lemon?

If you are planning to fry Jerusalem artichokes, peel and slice them, then fry them straight away. If they are young and fresh, it is unnecessary to parboil them first.

## Cooking

French cooks really know how to make the most of both Jerusalem and Chinese artichokes, putting them to good use in soufflés, gratins, sautés and daubes. A classic is Escoffier's Palestine soup, a rich milky purée made especially delicious by the addition of a handful of toasted chopped hazelnuts – the delicate flavours were made for each other.

The tubers blend well with a number of ingredients, but strong flavours tend to overwhelm their subtlety. They go well with potatoes – combine the two in a purée with butter, parsley and milk, or in a creamy gratin as an accompaniment to lamb or beef. You can also sit peeled whole artichokes round the roast for the last half hour of cooking. They will soften up nicely, agreeably absorbing the delicious juices from the meat. Another excellent dish is an autumn stir-fry of thickly sliced Jerusalem artichokes and field mushrooms, flavoured with garlic and plenty of sage.

Accompanied by robust, well-flavoured salad leaves, a mustardy dressing and maybe some crisp bacon nuggets or sweet meaty prawns, Jerusalem and Chinese artichokes make a great winter salad (try the Three-Artichoke Salad on page 158). Some people like them raw, but not everyone's digestive system is up to that; we prefer them lightly boiled or steamed.

## Nutrition

Jerusalem artichokes provide useful amounts of dietary fibre and relatively high levels of potassium and iron. They are also a good source of folate, often on the low side in roots and tubers.

## Varieties

Despite a reputation for knobbliness, there are smooth-skinned varieties of Jerusalem artichoke. A popular one is the elongated French variety '**Fuseau**'. Some of the samples we tried had a waxy texture – good in a cooked salad – and a sweet intense artichoke flavour. Tubers from another plot, however, were less sweet and not so well flavoured.

A number of named varieties are listed in the seed catalogues, but it is also worth growing the tubers of a favourite bought from the greengrocer, since there is a good chance that they will 'take'. We tried a smooth-skinned **red Jerusalem artichoke**, originally bought in a French market and grown by friends. It had a dense flesh that took longer to cook than 'Fuseau' and was not as sweet or as strongly flavoured, but still made a very good soup.

Although our home-grown **Chinese artichokes** had a nice firm texture, they were mild and unremarkable in flavour. However, samples bought from shops both in London and in Paris had an intense earthy flavour and crisp texture. They were good lightly steamed and served with butter and chopped parsley or chives.

## Three-Artichoke Salad with Chicory & Bacon▸

**Serves 4 as a starter or light meal**

*2 lemons*
*450 g/1 lb Jerusalem artichokes*
*2 garlic cloves, peeled*
*2–3 sprigs of thyme*
*115 g/4 oz Chinese artichokes (crosnes)*
*6 baby globe artichokes*
*extra-virgin sunflower oil or light olive oil, for frying*
*1 fat head of Witloof chicory*
*4 good handfuls of lamb's lettuce*
*6 rashers of streaky bacon, grilled until crisp*
*a few chives, snipped into longish lengths*
*a few sprigs of salad burnet or flat-leaved parsley*

**For the mustard dressing:**
*½ teaspoon English mustard powder*
*2 teaspoons Dijon mustard*
*½ teaspoon sugar*
*2 teaspoons white wine vinegar*
*salt and freshly ground black pepper*
*7 tablespoons extra-virgin sunflower oil or light olive oil*

**This is inspired by a dish served at The Ivy restaurant in London. If you don't have any Chinese artichokes, throw in a couple of extra Jerusalem artichokes. Lazy cooks can use bottled baby globe artichokes instead of fresh ones.**

**1** Have ready in a saucepan 1 litre/1¾ pints of water into which you have squeezed the lemons. Peel the Jerusalem artichokes and put them in the pan with the garlic and thyme. Bring to the boil, then simmer for 10–12 minutes, until they feel barely tender when prodded with a knife.
**2** Nip the ends off the Chinese artichokes if they look dry. Add them to the pan when the Jerusalem artichokes are nearly ready. Bring back to the boil and simmer for 1–2 minutes until just tender. Take the pan from the heat and leave the artichokes to cool in the liquid.
**3** Make the dressing: combine the mustards, sugar, vinegar and seasoning. Whisk in the oil in a thin stream until thick.
**4** Drain the cooked artichokes, discarding the liquid. Slice the Jerusalem artichokes and put them in a shallow dish with the Chinese artichokes. Pour over the dressing and leave to stand at room temperature.
**5** Meanwhile, prepare the globe artichokes and cut them into wedges (page 72). In a small frying pan, heat sunflower or olive oil to a depth of 2 cm/¾ inch. When it is very hot, but not smoking, tip in the artichoke wedges. Cover and fry over moderate to high heat for 5 minutes, until slightly crisp. Drain on paper towels.
**6** To assemble the salad, separate the chicory leaves and arrange on individual plates with the lamb's lettuce. Place the Jerusalem artichoke slices and globe artichoke wedges on top. Crumble the bacon and scatter over the salad with the Chinese artichokes and the herbs. Spoon the dressing remaining in the bowl over the salad.

## Jerusalem Artichokes with Baby Leeks, Pancetta & Sizzled Sage

**Serves 4 as a starter or side dish**

*450 g/1 lb Jerusalem artichokes*
*4–6 baby leeks, trimmed and sliced at an angle into 2.5 cm/1 inch pieces*
*55 g/2 oz pancetta or streaky bacon, diced*
*1 tablespoon lemon juice*
*sea salt flakes*
*freshly ground black pepper*
*4 tablespoons extra-virgin sunflower oil*
*handful of fresh sage, coarsely chopped*

**Jerusalem artichokes are related to the sunflower, and the rich satiny flavour of extra-virgin sunflower oil goes especially well with artichokes. If you have difficulty finding it, use light olive oil instead.**

**1** Bring a saucepan of salted water to the boil. Peel the artichokes and slice them thickly. Drop them into the water and simmer briskly for 15 minutes. Meanwhile, steam the leeks for 3 minutes, until they are just tender and still bright green. Brown the pancetta over moderate to high heat until crisp.
**2** Drain the artichokes and put them in a warmed serving bowl. Sprinkle with the lemon juice. Scatter the leeks and pancetta over the top and season with sea salt and plenty of pepper.
**3** Heat the oil in a small frying pan until very hot. Add the sage and sizzle for 30 seconds – it will crisp up as it cools. Pour the oil and sage over the vegetables and serve right away.

# SALSIFY, SCORZONERA & HAMBURG PARSLEY

Salsify *(Tragopogon porrifolius)* and scorzonera *(Scorzonera hispanica)* are sombre-looking vegetables. Salsify has dingy light-brown skin, covered with numerous fibrous whiskers, while scorzonera has bark-like brownish black skin with fewer whiskers. Both share the irritating habit of bleeding a milky latex when cut, making hands and pans difficult to clean. Nevertheless, they are worthwhile winter vegetables, with a distinctive subtle flavour.

Grown for its root rather than the leaves, Hamburg or turnip-rooted parsley *(Petroselinum crispum* var. *tuberosum)* is also dispiriting in appearance, rather resembling an albino carrot. However, a satisfying texture and flavour more than compensate for uninspiring looks. The fact that the root does not bleed latex is an added bonus.

## Family ties

Just like their human counterparts, plant families may be disparate and far-flung, but members tend to share broadly similar idiosyncrasies. For example, salsify and scorzonera belong to the Asteraceae family, which includes among its members the Jerusalem artichoke, infamous for its wind-producing inulin. True to form, inulin also features in both salsify and scorzonera.

The various members of the Apiaceae family, fennel, coriander and parsley – including Hamburg parsley – among them, are characterized by the essential oils that are responsible for their distinctive flavours and odours. The combination of essential oils characteristic of parsley permeates the whole plant, giving even the roots a recognizable 'parsley' flavour.

## Cultivation and harvesting

All three roots are sown from seed in the spring, ready for harvesting in autumn. They are cold-tolerant and can be left in the ground over winter, a practice which may improve their flavour by increasing sugar levels.

Although their roots are their chief asset, blanched salsify and scorzonera leaves are also worth eating. As a test we potted up a few salsify roots in late winter, covered the pots with black plastic and brought them inside. Two or three weeks later we had a harvest of tender, slightly bitter-sweet leaves that were delicious tossed in a tomato salad. Also good to eat in the spring are the leaves of the roots left in the ground over the winter. The leaves of Hamburg parsley are also edible, though repeated picking may reduce root growth.

## Buying and storing

In season from autumn to late spring, salsify and scorzonera are more likely to be found on market stalls and in specialist greengrocers than in the supermarkets. Unfortunately, Hamburg parsley is rarely seen.

Choose roots that are as whisker-free as possible, avoiding those that are broken or wizened or feel flabby. Treat them with care or they will start to bleed latex.

The roots keep for several weeks in a cool dark airy place, or the salad drawer of the fridge. Do not wash them until ready for use.

## Preparation

To avoid staining your hands, wear rubber gloves when preparing salsify and scorzonera. Quickly peel away all traces of skin and fibrous whiskers, immediately dropping the roots into water acidulated with plenty of lemon juice, to prevent discoloration. Alternatively, scorzonera can be cooked unpeeled, since the skin slips off easily enough afterwards.

Most recipes call for a preliminary boiling, before finishing the roots off in butter or cream. Put them in a pan and cover with cold acidulated water, or milk and a little water. Bring to the boil, then simmer for about 10 minutes, until only just tender. Watch the cooking time carefully, since the roots can quickly become soft and mushy. Drain, peel them if you have not already done so, and cut them into longish lengths.

## Cooking

Because of their subtlety, salsify and scorzonera are not at their best when in competition with stronger flavours. One of the tastiest and simplest ways of serving them is to toss lightly boiled roots in melted butter, a little lemon zest and plenty of chopped flat-leaved parsley. They are also good eaten cold, dressed with oil and lemon juice. Browned in oil and butter, they make excellent 'chips' – mouthwatering dipped in horseradish cream (page 162).

Hamburg parsley is delicious cooked in the same ways as carrots or celeriac. Cut it into small cubes and add to a meaty broth or casserole, or cut it into matchsticks for a Japanese-style stir-fry – it makes a good substitute for mooli. Alternatively, grate the root coarsely, toss it in lemon juice and serve as a winter salad; or mix with a creamy mustard dressing for *rémoulade*. If you can get hold of enough, Hamburg parsley will also make a richly flavoured mash, either on its own or mixed with potatoes.

## Nutrition

Like most root vegetables, salsify and scorzonera provide energy-rich carbohydrate as well as plenty of fibre. They contain potassium, calcium, magnesium and iron, and a useful amount of folate. Hamburg parsley is exceptionally rich in vitamin C .

## Varieties

Tasting trials were carried out in autumn and winter, with steamed or boiled roots.

'**Russian Giant**', a variety of scorzonera, is distinguished by its firm texture and mild flavour which, in the autumn tasting, was similar to that of artichokes. Since it does not disintegrate easily, this type of root is excellent for frying or grilling. It is delicious brushed with oil and cooked on a ridged cast-iron pan until slightly charred (page 162).

Also mildly flavoured is the salsify variety '**Sandwich Island**'. A winter tasting had a possible touch of 'oyster', giving credence to the sobriquet of 'oyster plant'. It was soft-textured and was good puréed with cream and plenty of chopped parsley.

We especially enjoyed the autumn harvest of **Hamburg parsley**. It had a smooth velvety texture and a clean, refreshing parsley flavour, with some celery in the background. Tasted in winter, the texture had become somewhat coarser, though the flavour was just as delicious.

## Pan-Fried Scorzonera with Horseradish Cream▸

**Serves 4 as a side dish**

*800 g/1¾ lb scorzonera*
*600 ml/1 pint milk*
*1 tablespoon olive oil*
*25 g/1 oz butter*
*2 garlic cloves, sliced*
*sea salt flakes*
*coarsely ground black pepper*
*3 tablespoons chopped flat-leaved parsley*
*lemon wedges, to garnish*

**For the horseradish cream:**

*3 tablespoons freshly grated horseradish*
*6 tablespoons thick cream*
*2 tablespoons Greek-style yoghurt*
*finely grated zest of ½ lemon*
*1 tablespoon lemon juice*
*1 teaspoon Dijon mustard*
*freshly ground black pepper*

**This is a delicious way of cooking scorzonera and goes very well with roast beef.**

1 To make the horseradish cream, combine all the ingredients in a small bowl and leave to stand at room temperature.

2 Peel the scorzonera and immediately place in a large saucepan with the milk. Add water if necessary just to cover. Bring to the boil and simmer for 10–12 minutes until just tender, then drain. Slice at an angle into 10 cm/4 inch pieces.

3 Heat the oil and butter in a frying pan until foaming. Add the scorzonera and cook for 2 minutes until beginning to brown. Add the garlic and cook for another minute or so. Drain on paper towels and transfer to a warm platter.

4 Sprinkle with sea salt flakes and several grindings of pepper. Shower with parsley and serve with lemon wedges and the horseradish cream.

## Grilled Salsify with Parmesan on Sage Farinata

**Serves 4 as a starter or light main course**

*450 g/1 lb salsify or scorzonera*
*juice of ½ lemon*
*olive oil, for brushing*
*70 g/2½ oz Parmesan, shaved into wafers*
*cherry tomatoes and frisée, to serve*
*sage sprigs, to garnish*

**For the farinata:**

*85 g/3 oz chickpea flour (gram flour)*
*½ teaspoon salt*
*300 ml/½ pint lukewarm water*
*1 tablespoon olive oil*
*small handful of fresh sage leaves, torn*
*coarsely ground black pepper*

**Farinata is a flat-bread made with chickpea flour. It has an earthy flavour which goes well with salsify.**

1 Make the farinata: preheat the oven to 240°C/475°F/gas 9. Sift the chickpea flour and salt into a mixing bowl. Gradually add half the water and whisk until you have a very smooth batter. Whisk in the rest of the water followed by the oil. Pour the batter into a lightly greased 22 cm/8½ inch square non-stick roasting tin. Sprinkle with the sage and black pepper to taste. Bake for 30–35 minutes until golden and crisp on top.

2 Meanwhile, peel the salsify and immediately place in a saucepan with the lemon juice and enough water to cover. Bring to the boil and simmer for 8–10 minutes until just tender, then drain. Slice into 9 cm/3½ inch pieces and brush with oil. Place on an oiled ridged stove-top grill. Cook over medium heat for 7–8 minutes, turning, until slightly browned.

3 Cut the farinata into four squares and place on a baking sheet. Arrange just over half the cheese on top, and melt under the grill for a few minutes. Place on serving plates with the tomatoes and frisée. Pile the salsify on top of the farinata, sprinkle with the remaining cheese and garnish with a sage sprig.

# OCA & MASHUA

The oca *(Oxalis tuberosa)* and the mashua *(Tropaeolum tuberosum)*, also known as añu, are ancient Andean tubers that remain important crops in that part of the world, which is home to hundreds of oca and mashua varieties. Outside the region neither has really caught on as a garden vegetable, though the oca has fared the better of the two. It is commercially produced in Mexico, Central America and New Zealand, where it is confusingly called a 'yam'.

Both tubers have odd flavours which initially take some getting used to. The oca has a reputation for being acidic or bitter, due to oxalic acid, though flavour may be improved by exposure to the sun after harvest, or storing for a few weeks before cooking. Glucosinolates make the mashua startlingly pungent, especially if tasted raw.

Both the oca and the mashua should be planted in the spring, in well-worked soil. They take their time, and do not produce tubers until the shorter days of late summer. They put up with frost and can remain in the ground until needed. If you want to harvest them all at once, however, they will keep for several months in a cool dark airy place, or in the fridge.

You are unlikely to find fresh tubers in the shops, although canned oca can sometimes be bought from stores catering for South American expats.

## Preparation and cooking

Ocas do not need peeling since the skin is very thin. It's worth peeling mashuas as their assertive flavour is concentrated in and near the skin.

Steaming preserves the beautiful skin colour of ocas. Cook them until just tender, then toss them in melted butter or serve them in a salad, perhaps mixed with new potatoes (opposite). Peruvian cooks serve them with a cheese and chilli sauce in *papas a la huancaina*, or with tomatoes, which go well with their slightly acidic flavour. Ocas are also good roasted in the oven. They become mellow and earthy, and turn a nice golden yellow, browning at the edges. Mashuas provide less scope for the cook, but we produced a reasonable dish by steaming and mashing them with cream and – the magic touch – freshly grated nutmeg.

## Nutrition

Ocas and mashuas should be appreciated for the nutrients they provide. An average serving of oca will provide almost the daily requirement of vitamin C, while mashuas contain nearly as much protein as an egg.

## Varieties

The ocas we grew were elongated, with distinctive cylindrical bulges. They had luminous light red skin and vibrant yellow-orange flesh. Though some varieties are purportedly bitter, the variety we grew had a pleasant, slightly tangy, tomato-like flavour.

Our mashuas were even more exotic – almost sinister – with tightly twisted bulges and lurid yellow skin flecked with dark red. Most were cone-shaped, though some were elongated and flattened. The flavour, though odd, was acceptable, especially after prolonged storage in the fridge.

# Peruvian Potato Salad▸

**Don't be put off if you can't find the 'potato' varieties used here. You can replace the ocas with waxy 'Pink Fir Apple' potatoes, and use 'Roseval' or 'Ratte' instead of purple potatoes. If making your own mayonnaise, use a light oil such as groundnut rather than extra-virgin olive oil, which is too strong.**

**Serves 4 as a light meal**

*450 g/1 lb ocas, unpeeled*
*280 g/10 oz purple-fleshed potatoes, such as 'Black Congo' or 'Truffe de Chine', unpeeled*
*1 head of radicchio or 'Treviso' chicory, leaves separated*
*4 hard-boiled eggs, quartered*
*1 small red onion,, cut into slivers*
*1 small red chilli, deseeded and cut into thin slivers*
*a few black olives*
*150 ml/¼ pint mayonnaise, preferably home-made*
*2 teaspoons dried red chilli flakes*

**1** Steam the ocas and purple potatoes in their skins for 10–15 minutes, until tender but not disintegrating. When cool enough to handle, peel the potatoes and slice the flesh crosswise. Leave the ocas unpeeled and slice lengthwise.

**2** Divide the radicchio or chicory leaves between individual plates and arrange the ocas, potatoes and egg quarters on top. Scatter with red onion and chilli slivers, and some olives. Add a dollop of mayonnaise and sprinkle this with chilli flakes.

# JICAMA & BURDOCK

Chosen because we like them rather than for their suitability as garden vegetables, jicama *(Pachyrhizus erosus)* and burdock *(Arctium lappa)* are among the exotics currently making an appearance on fashionable restaurant menus and in the larger supermarkets.

Apart from their curiosity value, the two have little in common. Jicama (pronounced 'heekama'), otherwise known as yam-bean, belongs to the legume family. It is a tropical climbing plant grown for its edible juicy tubers – its mature pods contain a toxic substance called rotenone, used as a fish poison and insecticide, and obviously should not be eaten.

Some varieties of jicama are shaped rather like a spinning top, others look like a four-cornered pouchy beret. The tubers have a papery light brown skin and ivory flesh, as crisp and juicy as a good eating apple. The flavour is subtly sweet and nutty, almost verging on bland. Brought to life by limes and chillies, jicama is usually associated with Mexican cooking, though it features equally in Southeast Asian cuisine.

In contrast to jicama's tropical requirements, burdock is better adapted to temperate regions. The leaves are edible but the plant is grown for its long tapering root, which can sometimes reach up to 1.2 m/4 feet in length. The root has a coarse brown skin, white crunchy flesh and a pungent, slightly bitter flavour.

Burdock is highly regarded in Japan, where it is known as *gobo*. Like the Jerusalem artichoke, it contains high levels of inulin, a polysaccharide with wind-inducing properties. Burdock also seems to have powerfully laxative properties, as we discovered after our recipe development sessions.

## Cultivation and harvesting

Jicama needs a long, warm growing season. Even though we nurtured it through the hottest of summers, both in a polytunnel and outside on a protected patio, we failed to produce full-size tubers. We therefore opt to use shop-bought jicama rather than grow it ourselves.

Burdock does well in cooler conditions. The seeds, which are sown in early spring, germinate best when exposed to the light, and should therefore be sprinkled over the surface of a compost-filled seed tray. Once germinated, the seedlings can be planted outside, ready for harvesting in summer and autumn. They do best in a sandy loam. Some determination and a sturdy spade or fork are needed to dig up the roots – those growing in heavy clay may be hard to prise from the soil.

## Buying and storing

Jicama is easily found in Chinese shops. Make sure the tubers are firm and fresh-looking, without any sign of bruises or mould. They store well, so it is worth buying two or three to experiment with. Wrapped in cling-film, the tubers will keep for a week or more in the salad drawer of the fridge.

Burdock is more elusive. The most likely sources are health food stores with a good fresh produce section, or Japanese food shops. The roots should be firm, not too long – about 30 cm/1 foot is ideal – and still encrusted with dirt. Left unwashed in a cool place, they can be stored for several weeks.

## Preparation

For maximum freshness, prepare jicama just before using. Remove the papery skin and inner fibrous layer with a small sharp knife. The flesh can then be sliced, cubed or grated as you like.

Burdock does not need peeling – indeed, it is the skin and the flesh immediately beneath it which has the most flavour. Scrub the roots under running water, then soak them in lightly acidulated water for at least half an hour to remove the bitterness. Slice them into thin strips, matchsticks or small cubes. Once cut, the flesh discolours very rapidly so immerse the pieces in yet more acidulated water as you work.

## Cooking

Jicama's refreshing juicy flesh is best appreciated raw. Try it instead of kohlrabi or mooli in salads; or cut it into cubes to serve with drinks, with a customary squeeze of lime, touch of salt and sprinkling of chilli powder. Though jicama can be steamed, boiled or baked like any other root or tuber, it is best stir-fried so that it keeps its crisp texture. Use it in the same way as water chestnuts.

Burdock can be boiled, fried or added to soups and casseroles just like any other root vegetable. We like it cut into matchsticks and stir-fried with carrots (page 168) – their sweetness offsets burdock's bitterness. It is also good in a spicy casserole with ginger, adzuki beans, rice, onions, sweet potato and carrots.

## Nutrition

Burdock contains useful nutrients, including calcium, magnesium and potassium, and a significant amount of dietary fibre. In comparison, jicama has a low nutritional value, but it does provide some vitamin C, especially when it is eaten raw.

## Varieties

Jicama seed is listed in only a few catalogues, usually under the generic name of 'jicama'. Burdock, too, is something of a rarity, and may be listed as either 'burdock' or 'gobo', though a few named varieties are also available.

Because of our heavy soil, we are unable to grow burdock. We tried some roots of '**Takinogawa Long**', the variety now most commonly listed in the catalogues, which were supplied by a colleague. They had a dense texture and an earthy, slightly bitter flavour that was improved by soaking in acidulated water.

Jicama

Burdock 'Takinogawa Long'

## Pico de Gallo▸

**Serves 6 as a starter**

*juice of 4 limes*
*1 teaspoon sea salt*
*1 small jicama*
*4 small oranges*
*½ red onion, thinly sliced into crescents*
*½ cucumber, thinly sliced*
*4 tablespoons extra-virgin sunflower oil*
*fresh coriander sprigs, to garnish*
*¼–½ teaspoon chimayo powdered chilli*

**There are different versions of this colourful Central Mexican dish. It is typically made with jicama, cucumber, succulent slices of fruit and fiery red chilli powder. If you don't have any jicama, use two or three crisp tart apples instead. You can use ordinary chilli powder instead of chimayo.**

**1** Combine the lime juice and salt in a small bowl. Peel the jicama, cut into quarters and slice thinly into segments. Cut the longer segments in half. Add to the bowl with the lime juice, toss and leave to stand.
**2** Using a sharp knife, cut a horizontal slice from the top and base of the oranges to expose the flesh. Remove the peel and pith by cutting downwards, following the contours of the fruit. Slice down between the flesh and membrane of each segment and ease out the flesh.
**3** Put the oranges, onion and cucumber in a serving bowl. Add the jicama and lime juice mixture, and the oil. Chill thoroughly.
**4** Toss again and garnish with coriander and powdered chilli just before serving. A nice touch is to dip one end of a few jicama pieces into the bright red chilli powder.

## Burdock & Carrots with Japanese Noodles

**Serves 2–3 as a side dish or light meal**

*1 burdock root about 30 cm/12 inches long, scrubbed*
*1 tablespoon extra-virgin sunflower oil*
*3–5 tablespoons stock or water*
*2 carrots, cut into matchstick strips*
*2 cm/¾ inch piece fresh ginger root, very finely chopped*
*½ teaspoon sea salt flakes*
*3 spring onions, green parts included, sliced*
*1 tablespoon tamari or shoyu*
*1 tablespoon mirin or dry sherry*
*dash of cayenne pepper*
*2 tablespoons chopped garlic chives*
*pinch of sesame seeds, to garnish*
*a few drops of toasted sesame oil*
*soba or udon noodles, to serve*

**This is adapted from John and Jan Belleme's excellent book *Culinary Treasures of Japan*. The sweet and warmish flavours of carrots and ginger counteract the slight bitterness of burdock. Earthy and mellow, soba or udon noodles are the perfect accompaniment.**

**1** Cut the burdock into 5 cm/2 inch matchstick strips, immersing them in acidulated water as you work.
**2** Heat the sunflower oil in a frying pan. Add the drained burdock and stir-fry over medium heat for 5 minutes. Moisten with stock or water, then cover and cook over very low heat for 15 minutes, until the burdock is almost tender. Make sure it does not stick to the pan.
**3** Add the carrots and ginger, with a bit more stock or oil if necessary. Season with salt and cook, covered, for another 7–10 minutes. When the liquid has nearly evaporated, add the spring onions, tamari, mirin, cayenne and garlic chives. Stir for another minute then sprinkle with sesame seeds and a few drops of sesame oil. Serve with the noodles.

# POTATOES

Cheap and ubiquitous though they may be, potatoes *(Solanum tuberosum)* are a vegetable to be reckoned with. There are over one hundred varieties available to gardeners, and an ever-increasing number in the supermarkets, including worthy old-fashioned varieties, new varieties and many that are organically grown.

## Texture and colour

Though on a superficial acquaintance potatoes may all seem pretty much alike, there are actually wide differences between varieties. Texture, for example, varies according to the amount of dry matter in the flesh. At one end of the continuum are the waxy types with relatively low levels of dry matter. Since they remain intact after cooking, they are perfect for boiling and for salads. At the other end are the floury sorts that have a high concentration of dry matter. These varieties tend to break up when boiled, but most are good for baking, roasting and chipping. About midway between waxy and floury come the general-purpose types.

While it is true that many potatoes are monotonously brown-skinned, some varieties have rosy pink or dramatic red and purple skins produced by anthocyanin pigments. Differences in colour are more than skin deep, however: they extend to the flesh, where carotenes create shades of ivory, cream and buttery yellow, and anthocyanins induce sumptuous swirling purples.

## Bitterness and greening

Potato tubers have the unpleasant habit of accumulating the bitter and potentially toxic glycoalkaloids solanine and chaconine. Usually the levels are too low to detect and rarely cause any harm. However, concentrations increase with exposure to light and can, if they become too high, render the tubers unfit to eat. A useful sign that glycoalkaloids are accumulating is greening of the skin. This is due to chlorophyll pigments that are, like the glycoalkaloids, made in the presence of light – the two processes occur simultaneously. The risk of bitterness and greening can be reduced by following the example of scrupulous gardeners who ridge up their crop to protect the tubers from light and then continue the blanching ritual by storing them in the dark.

## Cultivation

In the language of potato growing, the use of the word seed is somewhat muddled. In one sense, the term refers to true seed sexually produced when pollen, the plant equivalent of sperm, fertilizes eggs in the flower. In the case of the potato, these true seeds reside in the small, green, tomato-like fruits produced by the flowers and theoretically can be used to grow new plants. In sensible gardening practice, though, the use of true seed is not recommended since it results in small tubers and reduced yields.

For the best crops, commercially produced 'seed' potatoes should be planted. These 'seeds' are actually tubers harvested from plants produced from other tubers. Since tubers are specialized stems, their use for propagation is equivalent to taking stem cuttings from a geranium; the process is vegetative rather than sexual. It is also a type of cloning, and the seed tubers of a particular variety are exact genetic copies of one another.

## Harvesting

Potatoes conveniently fall into three categories, depending on when they mature. First earlies are ready in early summer. Their yields are relatively low, and they are eaten fresh instead of stored. 'New' potatoes are generally harvested from first early varieties before the skin has set. Second earlies can be harvested from mid-summer to late summer, and though they are also used fresh, some varieties store very well. Main-crop potatoes are the highest yielding and are harvested in the autumn.

## Buying and storing

To increase your chances of procuring flavoursome potatoes, buy them loose and unwashed (soil acts as a protective barrier) from an organic supplier or good greengrocer. Don't buy potatoes that feel spongy, are sprouting or are displaying green patches – all signs of poor storage.

If your potatoes were sold in a sealed plastic bag, remove them right away, otherwise condensation could cause rotting. If you are fortunate enough to have a cellar or frost-proof shed, you can keep potatoes for months, provided the environment is dry and well ventilated. Otherwise keep them in the coolest place possible, but preferably not in the fridge since the low temperature makes them unpleasantly sweet. If you have no choice but to keep potatoes in the fridge, follow the commercial practice of 'reconditioning' them by bringing them to room temperature two or three days before use. The sugars built up by chilling will then revert to starch.

## Preparation

Scrub potatoes well before cooking and remove any eyes that may have begun to sprout. Small patches of green can be cut away, but potatoes with an all-over greenish tinge are best thrown out.

Where possible, cook potatoes in their skins. Fewer nutrients will be lost, since these are concentrated just under the skin; you will lose less flesh, since the skin can be peeled away in a single layer after cooking; and floury types will keep their shape better.

For baked potatoes with a nice crispy skin, dry them well after scrubbing and prick them with a fork to allow steam to escape. On no account wrap them in foil, as the skin will simply steam rather than crisping.

To prepare potatoes for roasting, peel them and cut them into even-sized chunks with sharp edges – the edges become crispy as they roast. Put them in a pan of cold water, bring to the boil and simmer for 5–8 minutes. Drain and allow them to dry for a few minutes. Before tipping the potatoes into hot fat, bounce them up and down in the colander until they look slightly shaggy: they will then develop a good crisp coating when roasted.

To make perfect chips, use a fairly floury variety and prepare them in two stages – the first cooks them, the second crisps them. .Deep-fry them in oil preheated to 190°C/375°F for about 5 minutes until pale golden, and drain on crumpled paper towels. Then increase the oil temperature to 200°C/400°F, immerse the chips in small batches and fry for 2–3 minutes until crisp.

## Cooking

Potatoes are endlessly versatile and will good-naturedly team with a remarkable range of ingredients. They are also cosmopolitan, widely used not only in the cooking of South America, their country of origin, but in countries as disparate as India and Sweden, Turkey and Tunisia, Scotland and Spain.

Whether huge bakers or posh pebble-sized earlies, potatoes never fail to please as an accompaniment. They also form the basis of countless traditional dishes – bubbling creamy gratins, crisp-fried rösti, colcannon, hasselbacks and hash browns, to name but a few. Potatoes make good salads and snacks, they add substance to soups and their unique starchiness thickens casserole juices. Mashed until creamy, potatoes make excellent dumplings, such as plump little gnocchi, and can be used in pastries, breads and scones.

The deferential nature of potatoes makes them the perfect match for strident ingredients such as chillies, garlic, potent spices and strongly flavoured cheeses. Try making a massive omelette with crisp-fried new potatoes, onions cooked until meltingly soft, chopped fresh chilli, Parmesan and fresh herbs. Mix the whole lot with 6–8 lightly beaten eggs and tip it into a frying pan the right size to give you a really thick omelette. Cover and cook gently until the eggs are nearly set, then finish off the top with a blast under the grill.

Black Congo

'Charlotte'

Other favourites are Patatas Bravas (page 176) – bliss with a glass of Rioja or chilled dry sherry; a Nepalese potato salad dressed with chillies, toasted sesame seeds, fresh coriander and sesame oil; and Leslie Forbes's stunning Bombay Hash Browns, numbingly hot but curiously sweet and nutty at the same time: fry grated and dried potato with cumin, ginger, muscovado sugar, crushed cashews or peanuts, lime zest and juice and as many chopped fresh chillies as you like. Stir and toss until brown and crisp, then stir in creamed coconut and chopped fresh coriander.

We like chunky, but not lumpy, mashed potatoes. Instead of the usual butter and milk, add garlic and olive oil. Alternatively, try soured cream, dill and lemon zest, or anchovies, or a mash made golden with saffron threads, or Horseradish Mash with Sausages & Caramelized Onions (page 176). The possibilities are infinite.

## Nutrition

An average-sized freshly dug potato contains a sizeable dose of vitamin C – in the case of new potatoes almost the entire daily requirement. However, levels fall quite sharply during long-term storage. Steaming rather than boiling retains more of the vitamin. Potatoes are an important source of energy-rich carbohydrate, and if you eat the skins you'll also be getting a good supply of fibre. Potatoes also contain minerals such as potassium and magnesium, as well as useful amounts of folate.

## Varieties

When tasting we focused on second-early and main-crop potatoes. The tubers were harvested in the autumn and stored in a shed for three months. For the first stage of our tasting trials we steamed all the samples. For the second stage we cooked some varieties in other ways, depending on what was appropriate for their texture. With the cook in mind, we have divided the varieties into two broad categories: floury and waxy.

Among the floury potatoes, '**Edzell Blue**' (not shown) was our firm favourite. It has luminous white flesh, blue-purple skin, and a meaty, sweet and complex flavour. It makes terrific mash, but was just a tad too dry and floury when baked in its jacket. Another blue-skinned type is '**Arran Victory**' It has a creamy white flesh, a dry mealy texture and a pleasant, though mild, flavour.

The white-fleshed American '**Russet Burbank**' (not shown) is a large, bold potato. It has little depth of flavour but it is ideal for making fast-food-style thin chips, since it browns up quickly without losing its soft texture.

The American variety '**Red Pontiac**' has a texture somewhere between floury and waxy, moist with an assertive flavour. It is extremely versatile, lending itself to baking, roasting and chipping. We used it to make deliciously brown and crispy Provençal Potatoes (page 175) and it is very good sliced and baked in a gratin, since it is soft but still holds its shape.

'**Kennebec**', another American variety, is brown-skinned and white-fleshed, with a moderately floury texture and a pleasing flavour. It makes an excellent jacket potato, with a moist fluffy texture which forks up nicely.

'**Record**' (not shown) reputedly has a high dry-matter content, but we found the texture neither floury nor waxy, and the flavour unremarkable. However, it makes perfect golden chips with a crisp surface and moist interior. It also slices paper-thin for home-made crisps.

Of the waxy types, '**Belle de Fontenay**' (not shown) is a classic salad potato with yellow flesh that has a strong flavour. The curiously knobbly '**Pink Fir Apple**' has yellow flesh distinguished by an earthy flavour, which goes well in a salad of olives, feta cheese and lightly cooked green beans (page 184).

'**Roseval**' is another pink-skinned, yellow-fleshed variety. Its texture is between waxy and floury, and it has a good, though mild, potato flavour. It is a versatile potato, perfect for salads and also for boiling, mashing, or dicing into soups where you want bits of vegetable to retain their shape.

Brown-skinned '**Charlotte**' has yellow, waxy, slightly sticky flesh, with a mild flavour. Reports on '**Linzer Delikatess**' claim that it is a waxy type, but we found it somewhat floury. The flavour, though not strong, is fresh and light.

We did not grow '**Congo**', but we tasted a potato sold under the name of **Black Congo**. This has dark purple skin and purple and white marbled flesh. The texture is something of an oddity – fairly waxy but with a floury quality. The flavour is mild, but good. Teamed with the orange-fleshed oca in Peruvian Potato Salad (page 165), it provided the right balance to the vibrancy of the oca.

## ◂ New Potato Pizza with Pancetta, Rocket & Rosemary

**Makes one 30 cm/12 inch pizza**

*280 g/10 oz small waxy new potatoes, scrubbed but unpeeled*
*2 tablespoons olive oil, plus more for the pan*
*2 garlic cloves, finely chopped*
*1½ tablespoons chopped fresh rosemary*
*salt and freshly ground black pepper*
*85 g/3 oz smoked cheese, grated*
*115 g/4 oz Emmental or Gruyère cheese*
*55 g/2 oz pancetta, cut into small pieces*
*8 black olives, stoned and halved*
*handful of rocket, to garnish*

**For the pizza dough:**

*225 g/8 oz unbleached strong plain flour*
*1 teaspoon easy-blend dried yeast (½ sachet)*
*1 teaspoon salt*
*1 tablespoon olive oil, plus more for the bowl*
*125–150ml/4–5 fl oz tepid water*

**Potatoes as a pizza topping sounds odd, but the waxy type work very well and are especially delicious with Emmental cheese and pancetta.**

**1** First make the pizza dough: sift the flour, yeast and salt into a bowl. Make a well in the centre and pour in the oil and water. Stir, drawing in the flour from around the edge, until a soft dough forms. Knead for at least 10 minutes, stretching it with the heel of your hand. When the dough is silky smooth and elastic, place it in a lightly oiled bowl, turning to coat. Cover with cling-film and leave in a warm place for 1½–2 hours, until doubled in size.

**2** Preheat the oven to 240°C/475°F/gas 9. Plunge the potatoes into boiling salted water and blanch for 3 minutes. Drain, allow to cool and cut into thin slices.

**3** Heat the oil in a large frying pan. Add the potatoes and fry for 3–4 minutes, until lightly browned, turning carefully. Add the garlic, 1 tablespoon of the rosemary, and salt and pepper.

**4** Roll out the dough to a 30 cm/12 inch round and place on a preheated pizza stone or in a lightly oiled perforated pizza pan. Fold over the edge to make a rim. Sprinkle with about two-thirds of the cheeses. Arrange the potatoes over the top, followed by the remaining cheese. Scatter over the pancetta, olives and remaining rosemary. Lightly brush the edge with oil.

**5** Bake for 15–20 minutes, until browned and bubbling. Scatter over the rocket and serve.

## Provençal Potatoes

**Serves 4 as a side dish**

*1 kg/2¼ lb even-sized main-crop potatoes, unpeeled*
*4–5 tablespoons olive oil*
*25 g/1 oz butter*
*2 onions, thinly sliced into crescents*
*2–4 large garlic cloves, thinly sliced*
*3 tablespoons chopped flat-leaved parsley*
*1 tablespoon chopped fresh mixed herbs, such as rosemary, thyme and sage*
*sea salt flakes*
*freshly ground black pepper*
*flat-leaved parsley sprigs, to garnish*

**Crisp-fried potato slices with golden onions, garlic and fragrant Mediterranean herbs are perfection with grilled fish or meat. You will need a fairly floury potato, such as 'Kennebec' or 'Maris Piper'.**

**1** Put the potatoes in a pan with enough salted water to cover. Bring to the boil, then simmer for 5 minutes, until barely tender. Remove the skin and slice the flesh into thickish rounds.

**2** While the potatoes are cooling, heat 1 tablespoon of the oil and half the butter in a large frying pan, preferably non-stick. Add the onions and fry over moderate to high heat, stirring, until they begin to colour. Add more oil, if necessary. Add the garlic and fry until that also changes colour. Tip the whole lot into a bowl, mix with 1 tablespoon of the parsley, half the mixed herbs, a sprinkling of sea salt and black pepper. Keep warm while you cook the potatoes.

**3** Heat the remaining oil and butter until sizzling hot. Add the potatoes in batches and fry, turning with tongs, until golden brown. Drain each batch on paper towels.

**4** When the final batch is golden, lower the heat and return the rest of the potatoes to the pan together with the onion mixture and the remaining herbs. Stir together, check the seasoning and cook until warmed through. Garnish with parsley sprigs and serve.

## Patatas Bravas ▸

**Serves 4 as a side dish or 8 as a snack with drinks**

*900 g/2 lb firm-fleshed potatoes, peeled and cut into 2 cm/¾ inch pieces*
*5 tablespoons olive oil*
*1 tablespoon tomato purée*
*2 large garlic cloves, crushed*
*1 teaspoon paprika*
*½ teaspoon cayenne pepper*
*salt*

**This is probably not the authentic method for this traditional Spanish *tapas*, but it's the way we like it. Use an allpurpose main-crop potato such as 'Wilja', 'Red Pontiac' or 'Roseval'. To serve as a snack, spear the potato pieces with cocktail sticks and dip them in a bowl of mayonnaise.**

1 Preheat the oven to 180°C/350°F/gas 4. Steam the potatoes over boiling water for 5 minutes until barely tender. Alternatively, microwave them. Drain and spread out to dry on a clean tea towel for a few moments, then put them in a large bowl.

2 Combine the remaining ingredients, pour over the potatoes, and toss gently to coat.

3 Put the potato cubes in a roasting tin, spreading them out well in a single layer. If you overcrowd them, they will steam instead of crisping up and browning. Scrape in any oily residue remaining in the bowl.

4 Roast in the preheated oven, turning every so often, for 25–30 minutes, until a rich gold in colour and slightly blackened at the edges. Tip into a warmed bowl and serve without delay.

## Horseradish Mash with Sausages & Caramelized Onions

**Serves 4**

*675 g/1½ lb good-quality sausages*

**For the caramelized onions:**

*4 tablespoons olive oil*
*3 red or yellow onions, halved and sliced*
*1 tablespoon sugar*
*salt and freshly ground black pepper*
*1 tablespoon balsamic vinegar*

**For the horseradish mash:**

*1 kg/2¼ lb even-sized floury potatoes, unpeeled*
*4 tablespoons finely grated fresh horseradish, or to taste*
*125 ml/4 fl oz soured cream, crème fraîche or smetana*

**This dish is an absolute winner — creamy pungent mash, plump sticky sausages and sweet crispy onions. Use Cumberland sausages or good-quality pork chipolatas. If you don't have fresh horseradish, use the bottled kind, but it won't be quite the same.**

1 To make the caramelized onions, preheat the oven to 190°C/375°F/gas 5 and brush a large baking tray with half the oil. Spread the onion slices in a single layer on the baking tray. Sprinkle with the sugar, season with salt and pepper, and dribble over the remaining oil. Roast in the preheated oven for about 45 minutes, stirring them every 15 minutes, until golden.

2 Mix the vinegar into the onions, spread them out and roast them for another 7-10 minutes, until caramelized.

3 While the onions are cooking, prick the sausages with a fork and arrange them in a non-stick roasting tin. Cook in the oven with the onions for 35–40 minutes, turning occasionally, until nicely brown and sticky-looking.

4 Meanwhile, put the potatoes in a large saucepan with enough salted water to cover. Bring to the boil, then reduce the heat a bit and boil gently for 15–20 minutes, until very tender. Drain and put them back in the pan. Cover with a cloth and leave to dry for 5 minutes. Peel away the skin, then mash until smooth. Stir in the horseradish and the soured cream, crême fraîche or smetana. Season to taste with salt and pepper.

5 Spoon the potatoes into a dish and scatter the onions on top. Serve with the sausages.

# PODS & SEEDS

GREEN BEANS FRESH SHELLING BEANS

PEAS SWEETCORN

# GREEN BEANS

Among the most widely cultivated vegetables, beans can be divided into two main categories: those with edible fleshy pods (green beans) and those of which only the seeds are eaten (fresh shelling beans). The pods of green beans divide naturally into halves. As they mature, the pod walls become increasingly fibrous and unpalatable, while tough strings develop where the halves join. Fortunately, plant breeders have been able to work their magic on both French and runner beans, eliminating the strings, reducing the fibre content and producing varieties that remain tender over a longer period.

## Nitrogen fixers

As members of the legume family, beans have a special relationship with bacteria of the genus *Rhizobium*. These bacteria are often found naturally in the soil, though if they are absent the seeds can be inoculated with them when sowing. As the seeds germinate, the bacteria find their way into the roots of the plant, forming tiny nodules. By a process known as nitrogen fixing, *Rhizobia* extract nitrogen from the air and pass it on to the plant, where it is converted to protein and other nitrogen-containing

compounds necessary for growth. The plant returns the favour by giving the bacteria life-sustaining carbohydrates. This symbiotic relationship not only benefits both plant and bacteria, but creates, in effect, miniature fertilizer factories that are essential to the maintenance of agriculture worldwide.

## Cultivation and harvesting

Different beans have different climatic requirements. Asparagus peas and French beans, for example, are frost-sensitive, although they do perfectly well out of doors. In cooler environments, runner beans do better than French beans. Yard-long beans need plenty of heat and thrive best in a greenhouse or polytunnel, while broad beans can tolerate plenty of frost.

The ideal time for harvesting is when the seeds are still small, the pod wall has not yet developed too much fibre and the texture is crisp and crunchy. Experience will teach you the best size at which to pick the variety you grow.

'Aramis'

## Buying and storing

Freshness is paramount when buying beans; any showing signs of limpness or wrinkling are definite rejects. Look for smooth, firm specimens that snap cleanly when bent. Runner beans should not be discoloured or coarse. Yard-long beans should be a bright green. Don't worry about brown flecks on yard-longs – these are part of the pigmentation and will disappear during cooking. Beans in prime condition will keep for 3–4 days in a plastic bag in the salad drawer of the fridge.

## Preparation

If your beans need stringing, snap off both ends, pulling them back along the length of the bean to remove the strings. If the beans are stringless, line them up a handful at a time and chop off the stem ends. We like to leave the slender growing tips of French beans intact, but they can be removed if you wish. Once trimmed, leave the beans whole if they seem tender enough; otherwise cut them into short lengths.

Yard-long beans are stringless: nip off the tops and tails. We like to serve them whole – after all, you don't chop up spaghetti.

## Cooking

Although beans should normally be cooked, we like to add a few slivers of raw young runner beans to a green salad, along with a few of their brilliant crimson flowers.

For maximum flavour we prefer to steam our beans. Steam for 5–10 minutes, depending on variety and your preferred degree of doneness. Otherwise drop them into a large pan of boiling salted water and cook for 3–10 minutes, drain thoroughly and sizzle in melted butter or oil just before serving. If you like, add some finely grated lemon zest, chopped fresh coriander or garlic chives.

Green beans team happily with tomatoes, onions and mushrooms. One of the best-ever combinations is a simple Middle Eastern dish of green beans, garlic and tomatoes in olive oil (page 183). Green beans also add colour and texture to vegetable stews – drop them in towards the end of cooking.

## Nutrition

Although not overflowing with carotenes in the same way as leafy vegetables, green beans nevertheless provide valuable amounts of these vital nutrients. Being mildly flavoured, they may be a more palatable source to people who have a problem with leafy greens. Green beans also contain reasonable levels of all the other essential vitamins and minerals, particularly folate and vitamin C.

## Varieties

French beans, yard-long beans and asparagus peas were grown in polytunnels, broad beans and runner beans outdoors. For the tasting trials, the beans were steamed until just tender.

### French beans

Like oriental greens, French beans (*Phaseolus vulgaris*) bear the burden of numerous colloquial names. These include snap beans, bobby beans, *haricots verts*, and, confusingly, green beans. Many people also refer to them as 'string' beans, a legacy from the days when stringless varieties were not universal.

The growth habits of French beans are eclectic. They include dwarf varieties developed for machine harvesting, climbing types that need trellising, and a continuum of sizes in between. We tasted a number of both bush and climbing varieties

'**Fortex**', a climbing variety, is round and green, and has a nice beany flavour and good texture, even when almost 30 cm/12 inches long. Dressed with a little lemon juice and olive oil, it makes a great salad. Another climber, '**Kwintus**', a long green flat bean, also has a good flavour, though it is not quite equal to 'Fortex'.

The super-thin *filet* type '**Aramis**', a bush variety, is one of the delicacies of summer. We are usually unwilling to invest the time and care necessary for harvesting very thin beans, but in this case the flavour makes it worthwhile.

The climbing '**Kentucky Blue**' is smaller and darker green than 'Fortex', and also has an obvious beany taste. It is brilliant in a salad with yellow beans, a few waxy new potatoes, gleaming black olives and snow-white feta cheese (page 184). Almost as well flavoured is the flat light green bush variety '**Atlanta**'.

Of the yellow-podded beans, the oval-shaped '**Mont d'Or**' (not shown), a bush variety, is well flavoured and makes a colourful addition to a bean salad or a dish of plainly cooked green beans. '**Golden Sands**', another bush variety, is not quite as tasty, and the long flat '**Goldmarie**', a climber, has a mild flavour.

We generally do not concern ourselves with purple beans, since, once cooked, they become green anyway. That said, the bush variety '**Royalty**' has a good flavour and is well worth growing.

### Runner beans

The flavour of runner beans *(Phaseolus coccineus)* is on the bland side, and some people find the texture coarse and unrefined. Even so they make a good side dish, perhaps with a richly flavoured meat casserole. Just toss them in butter with a generous seasoning of sea salt and black pepper. Alternatively, use their somewhat fibrous nature to advantage and braise them Indian-style with tomatoes, chillies and spices (page 184).

### Asparagus peas

The asparagus pea *(Tetragonolobus purpureus)* is not a pea at all, but a type of square-podded bean, sometimes known as the 'winged bean' because of its four wavy fins. These beans should be harvested when no more than 2.5 cm/1 inch long. They are well worth trying for their delicious flavour, which, as the name suggests, is similar to asparagus. Cook them simply: either steam them or lightly sauté them in butter. You can find asparagus peas in Thai shops.

### Yard-long beans

Usually found in shops catering to Asian communities, yard-long beans (*Vigna sesquipedalis*) are easily identified by their prodigious length – though, at 45 cm/18 inches, the ones we grew were not quite as long as their name implies. They have a delicious, faintly lemony flavour, with a tender but resilient texture once cooked. As such, they stand up to braising and stir-frying, and reheat well. Try them with water chestnuts, shiitake mushrooms and a dash of lime juice to bring out their citrusy flavour.

### Broad beans

Though normally thought of as shelling beans, broad beans (*Vicia faba)* are delicious eaten as pods, when they are no more than 5–8 cm/2–3 inches long. Briefly steam them or blanch them in boiling water for 30 seconds, until they are only just tender, then drain them and toss them in melted butter or a little olive oil. A sprinkling of summer savory is a nice touch.

## Green Beans with Tomatoes, Garlic & Olive Oil

**Serves 4 as a starter**

*350 g/12 oz French beans, trimmed*
*5 tablespoons olive oil*
*½ onion, finely chopped*
*3 large garlic cloves, sliced*
*400 g/14 oz can of Italian plum tomatoes, drained and chopped*
*salt and freshly ground black pepper*
*generous pinch of sumac powder or finely grated zest of ½ lemon*
*pitta bread, to serve*

**Green beans and tomatoes make a perfect culinary partnership. This Lebanese dish is particularly good. Sumac is a coarse red powder used as a condiment in the Middle East. It has a pleasantly sharp, lemony flavour.**

**1** Bring a large pan of water to the boil. Cut the beans into 5 cm/2 inch lengths and drop them into the boiling water. Blanch for 2 minutes, then drain.

**2** Heat the oil in a large saucepan. Add the onion and fry gently over moderate heat for 5 minutes. When the onion is soft but not coloured, add the garlic and fry for another minute or two. Stir in the beans and toss until glossy and heated through.

**3** Add the tomatoes and season with salt and pepper to taste. Cover and simmer for 15 minutes, until the tomatoes have disintegrated a little.

**4** Pour into a warmed serving dish and sprinkle with sumac or lemon zest. Serve warm or at room temperature with plenty of pitta bread to mop up the juices.

## Mixed Bean Salad with Lemon Thyme Dressing▸

**Serves 4 as a starter**

*280 g/10 oz tender green and yellow beans*
*12 new potatoes*
*4 handfuls of frisée, cut into bite-sized pieces*
*8 black olives*
*25 g/1 oz feta cheese, crumbled*
*lemon thyme sprigs, to garnish*

**For the lemon thyme dressing:**

*1½ teaspoons lemon juice*
*½ teaspoon Dijon mustard*
*2 teaspoons chopped lemon thyme*
*sea salt*
*freshly ground black pepper*
*6 tablespoons extra-virgin olive oil*

**For an attractive presentation use a mixture of beans – yellow, green, flat and round, such as 'Mont d'Or', 'Fortex' and 'Kentucky Blue'. To enjoy the beans at their best , serve them while they are still slightly warm. Use a waxy new potato such as 'Pink Fir Apple' or 'Belle de Fontenay'.**

1 Whisk together the dressing ingredients, then leave to stand to allow the flavours to develop.

2 Plunge the beans into a large pan of boiling water. Cook for 5–6 minutes, until just tender but still brightly coloured and crunchy.

3 Cook the potatoes for 7–10 minutes, until they are just tender.

4 When the beans and potatoes have cooled but are still slightly warm, divide the frisée between individual plates. Arrange the beans and potatoes attractively on top. Add the olives and crumbled cheese, and garnish with a sprig of lemon thyme.

5 Whisk the dressing again and spoon over the salad.

## Runner Beans & Celery Tarkari

**Serves 4 as a side dish**

*450 g/1 lb runner beans, trimmed and strings removed*
*4–5 tender celery stalks, trimmed*
*2 teaspoons coriander seeds*
*1 teaspoon cumin seeds*
*3 tablespoons groundnut oil*
*½ teaspoon mustard seeds*
*¼ teaspoon asafoetida (optional)*
*½ teaspoon ground turmeric*
*¼ teaspoon garam masala*
*½ teaspoon sugar*
*½ teaspoon salt*
*½–1 small green chilli, deseeded and finely chopped*
*200 g/7 oz can of chopped tomatoes*
*2 teaspoons lime juice*
*2 tablespoons coconut flakes, to garnish*

**Tarkari is a type of Indian dish in which vegetables are cooked by a mixture of frying and steaming. The heat is raised towards the end of cooking to bring out the aroma of the spices. The slightly fibrous nature of runner beans and celery stands up well to the technique, but you could use French beans instead.**

1 Slice the beans and celery across at an angle into 1 cm/½ inch pieces. Drop these into a pan of boiling water and blanch for 1–2 minutes, then drain.

2 Dry-fry the coriander and cumin seeds in a small heavy-based frying pan until fragrant. Immediately remove the spices from the pan and grind them to a powder in a blender.

3 Heat the oil in a large frying pan. Add the mustard seeds and asafoetida, if you are using it. When the seeds begin to pop, stir in the ground coriander and cumin mixture, the turmeric, garam masala, sugar and salt. Fry for 30 seconds, then add the beans, celery and chilli.

4 Cover and cook over low heat for 10 minutes. When almost tender, add the tomatoes, increase the heat and cook for 5 minutes more.

5 Stir in the lime juice and garnish with the coconut flakes just before serving.

# Yard-Long Beans with Mushrooms & Water Chestnuts

**Serves 4 with rice as a main course, or 6 as a side dish**

*450 g/1 lb yard-long beans*
*3 tablespoons groundnut oil*
*115 g/4 oz shiitake mushrooms, sliced*
*1 small red chilli, deseeded and thinly sliced*
*2.5 cm/1 inch piece of fresh ginger root, very finely chopped*
*225 g/8 oz can of water chestnuts, drained*
*2 tablespoons lime juice*
*1 tablespoon soy sauce*
*1 teaspoon salt*
*3 tablespoons chopped fresh coriander*

**Most recipes instruct you to chop these beans into short lengths, but if you've got yard-long beans why not show them off? Twirl them round your fork like spaghetti. You can use ordinary green beans instead, if you don't have long ones.**

1 Blanch the beans for 2 minutes in a large pan of boiling salted water. Drain and dry thoroughly. Cut them in half if you wish.

2 Heat the oil in a wok or large frying pan until almost smoking. Add the mushrooms, chilli and ginger, and stir-fry over high heat for 2 minutes.

3 Add the beans and water chestnuts and fry until heated through, continually lifting and turning the beans. Stir in the lime juice, soy sauce, salt and coriander, tossing the beans until well coated. Serve at once.

# FRESH SHELLING BEANS

Plants, like animals, have an insatiable urge to reproduce. Initiating the process, flower pollination stimulates an orderly chain of events that inexorably leads to the formation of new seeds. As part of the development process, stores of proteins and starches are gradually built up in the seeds, which in turn produce embryos capable of producing new plants, ready to repeat the cycle themselves. Beans allowed to complete the reproductive cycle produce hard dry, but edible, seeds known as pulses. However, the cycle can be interrupted by harvesting the seeds while still moist and tender, and eating them as fresh shelling beans.

## Cultivation and harvesting

Broad beans are probably the most familiar fresh shelling bean. In a temperate climate they are a gardener's joy: uncomplaining and easy to grow. An autumn or spring sowing produces one of the first early summer crops, warmly welcomed by deprived devotees surfeited with a sulphur-laden winter routine of cabbage, kale and cauliflower. It is essential to pick broad beans young, while the seeds are still succulent and sweet. Otherwise the skin becomes tough and bitter, and must be removed.

There are others among the fresh shelling types including flageolet and kidney beans, which are equally easy to grow. Part of the attractions of these beans is the patterns and colours of the pods and seeds. In some cases they resemble beautifully polished pebbles – the pale pink borlotto kidney bean with its magenta streaks, for example. Others, like the light green flageolet 'Chevrier Vert', call to mind gentle watercolours flooded with subtle streaks and swirls. These types are summer crops, requiring a gardener with a finely tuned sense of timing. If picked prematurely, the seeds are too soft and insipid; left too late, they become dense and pulse-like. The goal is somewhere between the two, when the beans are firm but still moist.

Although they are usually grown for their pods, runner beans can be harvested later in the season, when the seeds have grown to full size but are not completely dry. At this stage the pods are too tough and stringy to be edible, but the  seeds can be surprisingly delicious.

## Buying and storing

Broad beans make an appearance in the shops from late spring onwards. The best are those bought early in the season, before they become oversized. Look for smooth, evenly shaped pods no more than 12.5–15 cm/5–6 inches long. Reject any that are limp or pockmarked. They will keep in the fridge for about three days but, once shelled, should be cooked within an hour or two.

While the late-summer harvest of fresh borlotti, flageolet and cannellini beans has long been enjoyed in southern Europe and the US, their appearance in Britain is relatively recent. The best sources are the larger supermarkets and Spanish or Italian shops. The state of the pods is a useful clue to the condition of the beans within. The signs of readiness are the opposite to those of edible-podded beans and broad beans: the pods should look tough, leathery and lumpy – a sign that the beans inside are plump and ready to be eaten. Like beans with edible pods, however, they should not be discoloured or show signs of decay.

Unpodded beans should be loosely packed in a well-ventilated plastic bag and kept in the fridge. If bought in good condition, they will keep for nearly a week, but watch out for signs of mould. Once shelled, the beans should be stored in a sealed container, refrigerated and used within twenty-four hours.

## Preparation

Broad beans are easy to shell, although you might want to wear rubber gloves as your hands can get stained black. Extricating other types from their pods, however, is a labour of love. Some pods are tough, tightly enclosing the beans inside like protective parents. You will have to use your thumbnail to open up the seams before prising out the beans.

Some varieties benefit from pre-cooking before being incorporated in another dish. Simmer them in water for 5 minutes, until they are barely tender, drain them and proceed with the recipe.

Unless they are very young and tender, broad beans need a 2-minute plunge in boiling water, followed by the removal of their tough outer skin. It slides off easily enough, revealing the brilliant green embryo inside.

Haricot borlotto 'Lingua di Fuoco'
Haricot 'Pea Bean'
Broad bean 'Stereo'
Runner bean 'Czar'

## Cooking

Fresh shelling beans are always eaten cooked. They are used in the same way as dried beans – in soups, salads and casseroles, or puréed – but have the obvious advantage of not needing soaking. They also cook much more quickly, sometimes in as little as 10 minutes, depending on the variety. Watch them carefully, since they quickly become mushy and lose their shape.

Young broad beans are perfect briefly steamed or boiled for a minute or two and served tossed with melted butter, sea salt flakes and, if you like, some snipped garlic chives or summer savory. Even shop-bought frozen beans are delicious this way. Just as mouthwatering are barely cooked broad beans sautéed in butter with spring onion and crisp bacon lardons; or heated through with plenty of cream, a generous amount of black pepper and served with pasta shapes. Large broad beans are not without redemption, even if mealy. Once the skins have been removed, they make a tasty purée mixed with plenty of butter and parsley.

Other types of fresh shelling beans, such as borlotti or cannellini beans, can be served in the same way as broad beans, although they will need longer cooking. They are good tossed in olive oil instead of butter, with crushed garlic, chopped fresh rosemary, thyme or epazote; or you can leave them to cool to room temperature and serve them as a salad, with chopped spring onion and perhaps some flaked tuna or sliced hard-boiled egg. If your beans are a bit bland, add a bouquet garni to the cooking water, along with some diced celery, leek and carrot for extra flavour.

## Nutrition

Fresh shelling beans are a powerhouse of nutrients – they are, after all, seeds from which the embryonic new plant develops and, as such, contain all the protein and complex sugars needed for growth. Compared to other vegetables, therefore, their protein levels are relatively high. That said, almost all beans lack one or more of the essential amino acids that are found in protein from animal sources. To produce 'complete' protein, fresh shelling beans should be eaten with grains, nuts and dairy products, which provide the missing amino acids.

Fresh shelling beans are an important source of slow-release carbohydrate, which stabilizes blood sugar levels. They are also extremely high in pectin – a form of soluble fibre which is thought to reduce blood cholesterol. They also contain insoluble fibre, which helps ease the passage of food through the gut, possibly reducing the risk of bowel cancer. Their fibre content is one of the reasons why shelling beans, fresh or dried, cause flatulence, especially if you are not used to eating them. Another reason is that they contain complex sugars known as oligosaccharides, which are bonded together in such a way that digestive enzymes cannot break them down. The sugars remain undigested until they reach the large intestine, where the bacteria present get to work on them, producing those unwelcome gases in the process.

Fresh shelling beans contain most of the B vitamins, including folate, as well as valuable amounts of calcium, iron, magnesium, zinc and potassium.

## Varieties

For our tasting trials we steamed the beans for 5–7 minutes, depending on variety.

### Broad beans

Variously known as fava beans, Windsor beans and sometimes even mistaken for the lima bean (a different bean altogether), broad beans *(Vicia faba)* are almost always harvested at the fresh shelling stage – in Europe and the US at least – and rarely left until dry. Most broad beans come in deceptively large pods that promise a bountiful harvest. In the best tradition of the braggart, however, they fail to deliver the goods, and the yield of each pod is disappointingly small for its size. Most produce seeds that are about 2.5 cm/1 inch long, but there are varieties with smaller seeds – the bean equivalent of *petit pois*. The seeds of **'Stereo'**, for example, are about half the size of a normal broad bean.

### Kidney beans

As the name suggests, these are plump, kidney-shaped beans. An excellent variety is the borlotto 'Lingua di Fuoco' *(Phaseolus vulgaris)*. It is fat and delicious, with a sweet earthy flavour and smooth texture. Combine kidney beans with broad beans and spelt *(farro)* in a hearty soup (page 192), or mix them in a salad (below) with other fresh shelling beans and power-packed herbs such as rosemary or epazote and a dressing of dark amber pumpkin seed oil.

Also worth trying are white **'Cannellini'** *(P. vulgaris)*. Although mildly flavoured, they add delicious body to soups and stews. The oddly named **'Pea Bean'** *(Phaseolus sp.)*, though not quite kidney-shaped, is another excellent variety with a sweet, full flavour and smooth texture.

### Flageolet beans

Pale green, small and slim, flageolets are delicacies in the bean world. We grew the mild-flavoured **'Chevrier Vert'** *(P. vulgaris*, not shown*)*, which looked and tasted delicious in a brick-red dish of chorizo, tomatoes and peppers (page 192).

### Runner beans

We grew the white-flowered variety **'Czar'** *(Phaseolus coccineus)*. The large white seeds can be tough-skinned and they need a relatively long cooking time, but they taste quite delicious, and are definitely to be recommended. There are other white-flowered varieities producing white seeds, and some with red flowers and dark seeds, which may well be worth trying.

## Bean Salad with Carrots and Pumpkin Seeds▸

**Serves 2–3 as a starter or light meal**

*225 g/8 oz fresh shelling beans*
*sea salt and freshly ground black pepper*
*1 teaspoon finely chopped epazote*
*1 tablespoon pumpkin seeds*
*1 spring onion, green part included, sliced*
*1 tablespoon chopped flat-leaved parsley*
*2 small carrots, thinly sliced*
*handful of 'Red Oak' lettuce leaves*

**For the pumpkin seed oil dressing:**

*2 teaspoons red wine vinegar*
*½ teaspoon Dijon mustard*
*sea salt and freshly ground black pepper*
*1½ tablespoons pumpkin seed oil*
*1½ tablespoons light olive oil*

**This salad looks lovely made with a mixture of differently coloured beans such as 'Lingua di Fuoco', 'Pea Bean' or 'Chevrier Vert'. You will need about 450 g/1 lb in the pod. If you can't get pumpkin seed oil use extra-virgin olive oil in place of that and the light olive oils. Use rosemary if you can't get epazote.**

1 Whisk the dressing ingredients until smooth and leave to stand while you cook the beans.
2 Bring a large pan of water to the boil, add the beans and cook for about 10 minutes, until tender but still holding their shape. Drain, tip into a bowl and pour the dressing over the beans while they are still warm. Season with salt and pepper, and stir in the epazote or rosemary. Leave to cool to room temperature, stirring occasionally.
3 When the beans are cool, mix in the pumpkin seeds, spring onion, parsley and the sliced carrots. Add more salt and pepper if necessary. Leave to stand at room temperature for at least 30 minutes to allow the flavours to develop.
4 To serve, arrange the salad leaves round the edge of a serving dish and pile the bean mixture in the middle.

# Mixed Bean Chilli with Chorizo ▸

**Serves 6**

*1 teaspoon cumin seeds*
*2 teaspoons coriander seeds*
*2 teaspoon dried oregano*
*675 g/1½ lb podded fresh shelling beans*
*3 tablespoons olive oil*
*2 onions, finely chopped*
*2 red peppers, deseeded and diced*
*2 garlic cloves, finely chopped*
*2–3 chillies, deseeded and finely chopped*
*225 g/8 oz chorizo sausage, thickly sliced*
*2 tablespoons tomato purée*
*400 g/14 oz can of chopped tomatoes*
*300 ml/½ pint meat or chicken stock*
*salt and freshly ground black pepper*
*4 tablespoons chopped fresh coriander*

**Fresh shelling beans cut down on the cooking time and impart a wonderful vibrant flavour and colour to the dish. Toasted crushed spices add richness and also help thicken the juices. The quantity of beans specified refers to the weight once the pods have been removed. You will need about twice the weight before podding. You can use drained canned beans if you like; in which case, omit the preliminary boiling. Stock should be home-made if possible.**

**1** Put the cumin and coriander seeds in a small heavy-based pan without any oil and dry-fry over moderate heat until fragrant. Sprinkle with the oregano, fry for a few seconds, then immediately remove from the pan. Using a pestle and mortar, lightly crush the mixture.

**2** Boil the beans for about 7 minutes until only just tender. Drain and set aside.

**3** Heat the oil in a large pan. Add the onions and the spice mixture and cook for a few minutes. Add the red peppers, garlic and chillies, and gently fry over medium heat for 5 minutes. When the peppers are soft, stir in the chorizo and tomato purée and cook for 2-3 minutes.

**4** Pour in the chopped tomatoes and stock. Season with salt and pepper bearing in mind the saltiness of the chorizo. Bring to the boil and add the beans. Simmer for 10 minutes.

**5** Just before serving, check the seasoning and stir in the coriander.

# Farro Soup with Borlotti and Broad Beans

**Serves 6**

*175 g/6 oz spelt, soaked for 2 hours*
*salt*
*2 tablespoons olive oil*
*1 tablespoon finely chopped fresh rosemary*
*85 g/3 oz pancetta, finely diced*
*½ red pepper, deseeded and finely diced*
*½ red onion, finely diced*
*2 celery stalks, finely diced*
*2 garlic cloves, finely chopped*
*salt and freshly ground black pepper*
*850 ml/1½ pints chicken stock*
*225 g/8 oz cooked borlotti beans*
*225 g/8 oz small broad beans*
*3 tablespoons chopped flat-leaved parsley*
*50 g/1¾ oz coarsely grated Parmesan cheese*

**Spelt, known as *farro* in Italy, is an ancient grain similar to pearl barley. You can buy it in health food shops and speciality food shops, or use brown rice instead. The soup is just as good made with canned borlotti beans and frozen broad beans, but it is worth using home-made stock.**

**1** Drain the spelt and put it in a saucepan with plenty of fresh water to cover. Add 1 teaspoon of salt, cover and bring to the boil. Reduce the heat and simmer for 35 minutes until tender, stirring now and again to prevent sticking. Drain, reserving the liquid.

**2** While the spelt is cooking, heat the oil with the rosemary in a large saucepan over medium heat. Gently fry the pancetta for 3-4 minutes until slightly crisp. Add the pepper, onion, celery and garlic, and cook for another 4-5 minutes until soft. Season with salt and pepper. Pour in the stock, then cover and simmer gently for 10 minutes.

**3** When the vegetables are tender, add the spelt and borlotti beans to the pot. Bring to the boil and simmer for 15 minutes. Add the broad beans and parsley, then simmer for 2-3 minutes more. The mixture should be quite thick and soupy, but you can thin it with the reserved spelt liquid if you wish.

**4** Ladle into bowls and sprinkle with the Parmesan cheese.

# PEAS

Reassuring and familiar, peas *(Pisum sativum)* are acceptable to the most recalcitrant of vegetable eaters. They show up in the smartest directors' dining rooms and the most basic of school canteens, to be shovelled up on whichever side of the fork is appropriate. The flavour is sweet, the mouth-feel is good and they don't need to be chopped or chewed.

The vegetable garden wouldn't be the same without peas. Their unique, sweet, grassy flavour is impatiently anticipated from the day the seeds go into the soil. Even before the first pods appear, we harvest the tender young shoots growing out from the vine – a crop often overlooked in the West, but enjoyed by Asian cooks.

Though shelling peas are eaten when they are young and sweet, they are classified according to their appearance when they are mature and dry. There are two types: wrinkle- and smooth-seeded. The wrinkle-seeded types are sweeter and, all things being equal, should be the first choice when choosing varieties for the garden.

## Unpalatable pods

As members of the legume family, peas produce seeds that are cocooned inside protective pods. The pods are usually inedible, because of a tough parchment-like membrane that lines the inner surface and is revealed only when the pod's tender outer layer is peeled away. The development of the parchment layer is under genetic control, and there are also membrane-free, edible-podded varieties. They include the flat-podded mangetouts, also called snow peas, as well as the rounder sugar snaps. Wonderfully sweet and succulent, both are much appreciated by cooks since preparation and wastage are minimal.

## Cultivation and harvesting

The choice of pea varieties is impressive, ranging from early to late-maturing, with both leafy and semi-leafless types and habits that range from dwarf- to tall-growing. They are hardy plants that thrive in cool temperatures. Smooth-seeded varieties are more cold-tolerant than wrinkle-seeded ones, and grow quite happily out of doors throughout the winter in milder areas. Peas also do well in polytunnels when sown either in autumn or in mid-winter. They develop more quickly than those grown outdoors, and are ready in spring when little else is available.

When it comes to harvesting, the best gardeners exercise stringent quality control, carefully picking the shelling varieties while the seeds are still sweet and tender. Mangetouts are ready when the seeds are barely discernible through the pod, while sugar snaps are best when their seeds reach the size of shelling peas.

Though pea shoots can be harvested from plants specially grown for the purpose, we harvest the tendrils with two or three pairs of leaves from plants grown for their pods, doing so in moderation so as not to reduce yields.

## Buying and storing

Look for shelling peas with plump, bright green, wrinkle-free pods. Choose crisp, flat mangetouts, avoiding those that are bumpy or limp. Sugar snaps should look crisp and fresh. Peas will keep for as long as a week in a plastic bag in the fridge, but watch out for mould at the stem end.

Pea shoots are occasionally found in oriental stores. If they are green and sprightly, they usually keep for up to three days in the fridge in a plastic box lined with damp paper towels.

## Preparation

When preparing shelling peas, bear in mind that 450 g/1 lb provides 225 g/8 oz – about enough for two. If you need to shell them in advance, do so no more than an hour or two before serving, and store them in the fridge.

To remove mangetout and sugar snap strings, break off each end and pull down the sides of the pod. Slice mangetouts at a sharp angle for stir-fries. Sugar snaps are best left whole – if you slice them, the seeds escape and risk being overcooked.

To prepare pea shoots, discard any excessively large leaves and tough stems, then rinse them briefly.

## Cooking

Though seemingly unworldly, peas are remarkably cosmopolitan. Oriental cooks use them in stir-fries and soups, Indians add them to curries and fritters, and Italians toss them into pasta and rice dishes, In Britain, peas are the classic accompaniment to roast duck, while in France they are traditionally served with lamb, veal and poultry.

When cooking any type of pea, keep it quick. Steam them, or cook them in a very little ready boiling salted water until just tender, then toss them in melted butter, with perhaps a sprinkle of a delicately flavoured herb such as lemon balm or chervil.

Shelling peas are delicious lightly braised with compact little lettuce hearts and new potatoes (page 196), or use them in a quick pasta sauce with snippets of Parma ham, basil and lots of cream – delicious with fresh tagliatelle.

Mangetouts add crunch and colour to stir-fries, and are good blanched briefly and added to a salad. Try them with an oriental mix of peppery Chinese leaves, crisp-fried chicken nuggets, spring onions and almonds, tossed with soy sauce and sesame oil.

Pea shoots need tender treatment. Enjoy them raw in a salad, sprinkled with a few drops of lemon juice and extra-virgin sunflower oil, steam them lightly, stir-fry them till just wilted, or use them to crown a stir-fry of shelling peas, sugar snaps and mangetouts (page 196).

## Nutrition

Peas are among the most nutritious vegetables – some comfort for parents of young children. They contain a significant amount of protein, and plenty of carbohydrate and dietary fibre. They are also rich in B vitamins and vitamin C, and well-endowed with minerals, including calcium, magnesium, iron and zinc.

## Varieties

Tastings were carried out on tunnel-grown peas sown in mid-winter and harvested in late spring. All the peas were steamed.

The seeds of '**Lincoln**' (not shown) and '**Feltham First**' (not shown) were tasted at various stages of maturity. 'Lincoln', a wrinkle-seeded variety, was invariably sweeter, livelier and more tender than the smooth-seeded 'Feltham First'. '**Waverex**' (not shown), a small-seeded *petit pois* variety, was sweet, tender and delicious, though its small pods and seeds make a meal somewhat labour intensive – it's easier to add a few raw seeds to a salad.

Of the edible-podded peas, the mangetout '**Oregon Sugar Pod**', a tall variety, needing support, has a nice green grassy flavour with a hint of sweetness. It makes a superb stir-fry, adding brilliant colour and crisp texture. '**Sugar Snap**' (not shown), another tall variety, has thick-walled fat pods that are sweet and full-flavoured, with earthy tones in the background. '**Sugar Ann**', another sugar snap, is less sweet but still delicious. They are all excellent braised with young carrots, tender asparagus tips and spring onions.

## Many-Pea Stir-Fry with Chicken & Rice Noodles▸

**Serves 4**

*450g/1 lb boneless, skinless chicken breasts*
*70g/2½ oz thin rice noodles*
*groundnut oil for deep-frying, plus 3 tablespoons*
*6 fat spring onions, sliced*
*1 cm/½ inch piece fresh ginger root, thinly sliced*
*175 g/6 oz sugar snaps*
*115 g/4 oz mangetout, sliced at an angle*
*140 g/5 oz shelled peas*
*40 g/1½ oz whole almonds with skin, halved lengthwise*
*55 g/2 oz beansprouts*
*handful of pea shoots (optional)*

**For the marinade:**

*1 tablespoon muscovado sugar*
*3 tablespoons warm water*
*1–2 small red chillies, deseeded and very finely chopped*
*1 tablespoon soy sauce*
*1 teaspoon nam pla fish sauce*
*3 tablespoons lime juice*

**A celebration of peas if ever there was one. Of course, you could use all sugar snaps or all mangetouts instead of a combination, but each adds a different texture and flavour to the dish. Pea shoots are an indulgence – lovely if you grow them, but not strictly essential.**

**1** Combine the marinade ingredients in a bowl, first dissolving the sugar in the warm water. Cut the chicken into bite-sized pieces and add to the marinade. Leave to stand while you prepare the rest of the ingredients.

**2** Pull the bundle of rice noodles apart, and break them into shorter lengths to make them easier to fry. Do this in a deep bowl so they don't fly all over your kitchen.

**3** In a large frying pan, heat enough oil to come to a depth of 5 mm/¼ inch. When the oil is very hot, throw in the noodles and fry for just a few seconds until puffed up. Be careful not to let them burn. Drain on paper towels and keep warm.

**4** Drain the oil and wipe out the pan with paper towels. Heat 3 tablespoons of clean oil over moderate to high heat. Fry the onions and ginger for 30 seconds, then add the chicken with the marinade and fry for about 5 minutes. When the liquid has reduced, add the three types of peas and fry for another 2–3 minutes. Don't let the peas overcook – they should remain bright green.

**5** Add the almonds, the beansprouts and, if you are using them, the pea shoots, and stir for a few seconds, until they are heated through. Divide the rice noodles between individual plates and serve with the stir-fry.

## Braised Peas & 'Little Gem' Hearts with New Potatoes

**Serves 2 as a light meal**

*350 g/12 oz small new potatoes*
*sea salt flakes*
*freshly ground black pepper*
*large knob of unsalted butter*
*4 fat spring onions, green parts included, thickly sliced*
*2 'Little Gem' lettuce hearts, quartered lengthwise*
*175 g/6 oz shelled peas*
*lemon balm, mint or chervil sprigs, to garnish*

**Freshly shelled peas, crunchy lettuce hearts and earthy new potatoes tossed in a generous amount of butter make great summer eating. Use the outer lettuce leaves in a salad.**

**1** Cook the potatoes in their skins until just tender. Drain, season to taste, and keep warm.

**2** Melt the butter in a frying pan. Add the spring onions and cook over medium heat until just soft. Throw in the lettuce and peas, along with a splash of water or stock. Cook for 1–2 minutes, until the lettuce is just tender but still brightly coloured and crisp.

**3** Add the potatoes to the pan. Check the seasoning and strew with herbs before rushing to the table. Not a dish to be kept waiting.

# SWEETCORN

.Every spring we have a long discussion about whether or not to grow sweetcorn *(Zea mays)*. If our decision were based on rationality we wouldn't ever grow it, since it is a tall crop that is not only greedy with its space requirements, but is also a laggardly cropper – it takes almost four months for the first ears to mature. Fortunately, our emotional side always wins out, and we end up planting sweetcorn every year. Any lingering misgivings are dispelled when we bite into the first ears of the summer, relishing the sweet flavours that explode in the mouth like tiny firecrackers.

## Sugar and starch

The pleasure of eating sweetcorn is primarily related to the amount of sugars present in the kernels and the rate at which these sugars are converted into starch. $F_1$ hybrids of sweetcorn conveniently fall into three main types, distinguished by genetic differences in their sugar levels. The least sweet are known as 'normal sugary'. They have a sugar content as low as 8–10 per cent, which, even under refrigeration, converts to starch in a day or two. The scientists have been tinkering, however, and have come up with hybrids to cater for those with a sweeter tooth. Types known as 'supersweets' contain up to 50 per cent sugar, which can take over a week to convert to starch. Another group, 'sugary-enhanced', falls between the 'supersweets' and 'normal sugaries'.

## Corn terminology

It is worth taking a moment to clarify the terms 'corn' and 'maize'. 'Corn' may be used in a general sense to describe the staple crop of a region – wheat in England or oats in Scotland, for example. The first European settlers in America, seeing that maize was the staple grown by the natives, called it 'Indian corn'. In modern America, 'corn' is still commonly used to refer to maize. In Britain, however, 'corn-on-the-cob' or 'sweetcorn' is used to differentiate this from 'corn' (wheat), the staple, while 'maize' refers to cattle fodder.

Sweetcorn is one of five main categories of corn. It is eaten while young and fresh, while other harder forms of corn – dent or field corn, flint or Indian corn, flour corn and popcorn – are normally processed into cornmeal and various corn-based products.

## Promoting pollination

The corn plant produces two types of flowers: a male tassel that develops at the top of the plant and produces pollen; and a female ear that grows from the stalk, bearing numerous strands of silk. Each strand leads to the development of only a single kernel, and most of the silks must be pollinated if plump, well-packed ears are to be harvested by the end of the season.

Occasionally, sweetcorn is self-pollinated when the pollen falls from the tassel on to the ear of the same plant. Cross-pollination, however, is much more common since the wind readily moves pollen between plants. To improve the chances of pollination, sweetcorn should be grown in a block of at least four rows, and never in a single row. The only exception to the rule is baby corn, which is harvested when the silks first begin to emerge and does not need pollinating. Because of sweetcorn's tendency to cross-pollinate, quality may be sacrificed when varieties of more than one type are grown close to each other. For example, crosses between sweetcorn and field corn, popcorn or ornamental corn will produce starchy kernels. The problem is not so serious when cross-pollination takes place between normal sugary and sugary-enhanced types. However, if pollination occurs between the supersweets and either the normal sugary or sugary-enhanced types, then the kernels produced will be tough and starchy.

## Harvesting

Depending on area, harvesting can take place from mid-summer to autumn. The ears should be ready once the silks have turned brown, but to make doubly sure spread apart the leafy husks and pierce one or two kernels with your thumbnail. If the juice that spurts out looks milky, the ear is ready. After three or four tries with the piercing technique, you will be able to judge readiness on appearance alone. Conversion from sugar to starch begins from the moment of harvest, so be ready to cook your corn right away.

## Buying and storing

If possible, buy sweetcorn from a greengrocer or farmers' market rather than a supermarket, as it is more likely to be freshly picked.

Choose ears that are completely enclosed in fresh green husks; if the husks are opened or removed the kernels dry out more quickly. Avoid ears with pale silks, indicating that the corn was picked too soon. Peel back a bit of husk and check that the kernels are plump and tightly packed in gleaming rows.

Because sweetcorn starts to lose sweetness from the moment of harvest, it is best eaten on the day of purchase. If this is not possible, wrap unhusked ears in damp paper towels and store in a plastic bag in the fridge. They will keep for up to two days, but will continue to lose sweetness. Even supersweets taste stale after a couple of days, although they do keep their sweetness longer.

## Preparation

Perhaps sweetcorn's single most irritating trait is the silks, which obstinately cling either to your fingers or to the kernels. A solution is cut off about 2 cm/¾ inch from the base rather than the tip of the ear. Loosen the husk a little at the bottom, working all round the cob, then grasp the ends firmly and strip it back from bottom to tip. The silks should come away at the same time. If this fails, you will still have a tassel of silks intact at the tip of the ear, which you can pull off in one go.

Another solution is to leave the husks in place until after cooking. You'll find that moist or wet silks come away more easily than dry ones.

If you need to remove the kernels before cooking – for a soup or casserole for example – dehusk the ears and cut them in half crossways to make them easier to handle. Stand the pieces upright with the narrow end at the top and shave away the kernels with a small sharp knife, cutting close to the cob. Break up any clumps with your fingers.

## Cooking

Sweetcorn is almost symbolic of homely North American cooking. Fresh, frozen, canned or creamed, it is served on the cob, in soups and chowders, relishes and pickles, and in custards, puddings and breads. As far as Europe is concerned, it seems that Britain alone shares the American enthusiasm; other countries rarely serve it. Sweetcorn, particularly in the form of miniature ears, also crops up in Southeast Asia, where it is added to stir-fries, seafood dishes and soups.

For all sweetcorn's versatility, it is hard to improve on plainly boiled ears with melted butter, a scattering of sea salt flakes and perhaps some chopped mint or chives, or a dash of dried chilli flakes.

Barbecued sweetcorn is another gastronomic high. It is best to grill the ears in their husks rather than blackening the kernels directly over the coals, which makes them tough. Some people like to soak the ears in cold water for 30 minutes before cooking, but if your corn is fresh and juicy this may not be necessary. The kernels will be steamily moist and tender, blissfully infused both with the aroma of the barbecue and the husk. If you insist on a few authentic black bits, partially open the husks towards the end of cooking.

Try forgoing the butter for a change and serve your sweetcorn Mexican-style, sprinkled with lime juice, salt and coriander, or even epazote if you grow it (page 202). Alternatively, dribble over some good olive oil flavoured with garlic or chilli. Forget about dinky little corn holders – sweetcorn is an in-your-face job. Hold the cob in your fingers and suck on it until every last kernel has been devoured.

Once you've reached saturation point with boiled or barbecued corn, you may wish to consider other ideas. Sweetcorn has a real affinity with dairy products and chillies; dairy products seem to reinforce the creaminess, while chillies provide the perfect contrast to the sweetness. Seafood is good with corn, too, perhaps because of its inherent sweetness.

Our favourite combination is a deeply satisfying layered casserole of sweetcorn kernels, grilled tomato sauce, chillies and crisp-fried corn tortillas dripping with melted cheese (page 201). Barely cooked kernels are also good in a salsa with finely diced tomatoes, spring onion, chillies, lime juice and fresh coriander – perfect with grilled prawns or lobster.

A homesick Chilean friend taught us how to make *pastel de choclo* (corn pie), a favourite national dish something along the lines of shepherd's pie, but with a thick topping of puréed sweetcorn, milk and onion instead of mashed potato. This is ideal for a big family get-together since you need plenty of willing hands to prepare the huge amount of corn needed for the topping.

We would also like to put in a plea for succotash, a creamy mishmash of corn, beans and, it seems, anything else that happens to be around. This may cause the hearts of some Americans to sink since the dish was the archetype of bad cafeteria food in the 'fifties and 'sixties. Succotash is more than redeemed, however, when made with lightly cooked fresh sweetcorn kernels, tender broad beans, spring onions and snippets of crisp-fried bacon. We prefer to leave out the cream.

## Nutrition

Sweetcorn is the ideal summer snack for children since it is a good source of energy-rich carbohydrate. It also provides small but useful amounts of protein and B vitamins, needed for the release of energy from food, as well as vitamin C, potassium and magnesium.

## Varieties

The majority of sweetcorn varieties are $F_1$ hybrids. We hesitate to recommend specific ones, since sweetcorn seems to come and go with remarkable speed; what is in the catalogues one year may not be there by the next. However, there is an excellent choice and it is simply a matter of deciding on what appeals to you and what is suitable for growing in your area.

Unlike the colour carnival of peppers and tomatoes, the sweetcorn palette consists mainly of soothing yellows, milky whites and bicolour mixes of yellow and white. Gardeners and cooks alike often show strong preferences for a particular colour, but this is somewhat irrational since colour pigments probably have no effect on flavour.

We invariably grow the supersweets for their exceptional flavour and crisp, crunchy texture. They are superb plainly boiled or grilled in their husks and served with a herb butter. Even so, some supersweets can be too sweet and we occasionally long for the old-fashioned texture and flavour of the normal sugary and sugary-enhanced types. Unlike the low-starch supersweets, these contain fairly high levels of a specialized starch called phytoglycogen, which gives the kernels a creamy texture and a characteristically corny flavour. We prefer to use them for recipes calling for puréed corn, such as the deliciously creamy Sweetcorn, Green Pepper and Chill Chowder (page 202), and for corn breads.

Though there are specific baby corn varieties in the catalogues, any type of corn, including field corn, can be used if the cobs are harvested young enough. We sometimes get so impatient waiting for the first cobs to mature that we harvest them as baby corn.

# Corn & Tortilla Casserole

**Serves 4-6 as a main course**

*4 large ears of sweetcorn, with husks, or 450 g/1 lb frozen kernels, defrosted*
*625 g/1 lb 6 oz tomatoes*
*4 large mild green chillies, such as Anaheim*
*½ onion, chopped*
*2 garlic cloves, chopped*
*2 tablespoons groundnut oil, plus more for frying*
*2 teaspoons cumin seeds, toasted and crushed*
*salt and freshly ground black pepper*
*12 small corn tortillas, torn into 2.5 cm/1 inch strips*
*225 g/8 oz mild Cheddar or Monterey Jack cheese, grated*
*1 tablespoon chopped fresh coriander*

**This is one of the most deeply satisfying dishes ever. It's even worth making with frozen sweetcorn and bottled enchilada sauce if you're short of time. However, it is essential to use corn tortillas — the wheat variety simply will not do. If you don't have any Anaheim chillies, use a large green pepper and a small green chilli instead.**

**1** Place the sweetcorn ears in their husks under a preheated very hot grill for about 10 minutes, turning occasionally, until just tender. Peel away the husks and silks, and slice the kernels from the cob (see Preparation, page 199).

**2** Place the tomatoes and chillies under the grill, turning occasionally, until slightly blackened. Remove the skin and seeds from the chillies and chop the flesh roughly. Put the tomatoes (including the blackened skin), onion and garlic in a food processor or blender and process to a chunky purée.

**3** Heat the 2 tablespoons of oil in a medium-sized frying pan, until almost smoking. Pour in the tomato mixture and stir over a fairly high heat for 5 minutes until slightly thickened. Add the crushed cumin seeds and season with salt and pepper to taste. Tip the mixture into a large bowl.

**4** Preheat the oven to 190°C/375°F/gas 5. Pour enough oil into a large frying pan to reach a depth of 5 mm/¼ inch. When it is almost smoking, toss in the tortilla strips in batches, and fry until just crisp. Drain on paper towels, then stir them into the tomato mixture until well coated.

**5** Arrange one-third of the tortilla strips over the bottom of a lightly greased 2 litre/3½ pint deep ovenproof dish. Sprinkle with half the chopped chillies, corn and cheese, and season with salt and pepper. Add another layer of tortilla strips, followed by the remaining corn and all but 6 tablespoonfuls of the cheese. Season again, add a final layer of tortilla strips and sprinkle with the remaining cheese.

**6** Bake in the preheated oven for 30–40 minutes, until bubbling and golden brown. Sprinkle with the coriander just before serving.

## Grilled Sweetcorn with Lime & Coriander▸

**Serves 4 as a starter or snack**

*4 sweetcorn ears, with husks*
*juice of 2 limes*
*1 tablespoon chopped fresh coriander or epazote*
*generous pinch of dried chilli flakes*
*sea salt flakes*

**For the chilli and lime butter (optional):**

*115 g/4 oz unsalted butter*
*2 tablespoons chopped fresh coriander or epazote*
*finely grated zest of 1 lime*
*1–2 teaspoons dried chilli flakes*

**The sharp clean taste of the lime and coriander – or epazote if you grow it – is perfect with the sweetness of the corn. If sweetcorn without butter cannot be contemplated, however, serve it with chilli and lime butter instead.**

**1** If you want to serve the corn with chilli and lime butter, mash all the ingredients together with a fork. Form into a cylinder and chill well before cutting across into slices.

**2** Place the corn ears in their husks under a preheated very hot grill or over hot coals for about 10 minutes, turning occasionally.

**3** Slit the husks and silks lengthwise and pull apart to expose the corn. Squeeze over the lime juice and sprinkle with chopped coriander or epazote, chilli flakes and sea salt. Eat straight from the husk. Alternatively, remove the husks and silks (see Preparation, page 199), chop the cobs into three or four pieces and put them in a serving bowl. Then add the seasonings.

## Sweetcorn, Green Pepper & Chilli Chowder

**Serves 4–6**

*4 sweetcorn ears, with husks, or 450 g/1 lb frozen kernels, defrosted*
*1 green pepper, halved and deseeded*
*1 small fleshy green chilli*
*400 ml/14 fl oz semi-skimmed milk*
*1 tablespoon sunflower or grapeseed oil*
*1 small onion, finely diced*
*2–3 sprigs of thyme*
*1 bay leaf*
*2–3 small potatoes, finely diced*
*600 ml/1 pint chicken stock or vegetable stock*
*salt and freshly ground black pepper*
*3 tablespoons chopped fresh coriander*

**Tasting mildly of chillies and made with freshly picked sweetcorn, this soup is out of this world. It's pretty good with frozen sweetcorn too. If you do use frozen kernels, there's no need to cook them before you purée them.**

**1** Place the sweetcorn ears, in their husks, and the green pepper, cut side down, in a pan under a very hot grill. Grill for 5 minutes, turning the corn occasionally, then add the chilli. Grill for another 10 minutes until the corn husks are browned and the pepper and chilli are beginning to blacken and blister.

**2** Peel away the husks and silks, and slice the kernels from the cob (see Preparation, page 199). Peel the pepper and chilli, remove the seeds from the chilli, and chop the flesh of both into small dice.

**3** Purée half the corn with the milk in a food processor or blender until reasonably smooth.

**4** Heat the oil in a saucepan and gently fry the onion with the thyme sprigs and bay leaf until the onion is translucent. Add the pepper, chilli, potatoes, puréed corn and the stock. Bring to the boil, then simmer gently for 15 minutes. Add the remaining corn and simmer for another 5 minutes.

**5** Fish out the herbs and season to taste with salt and pepper. Purée half the mixture, then return this to the soup in the pan. Reheat gently and stir in the coriander.

# SUPPLIERS

**Seeds were sourced in both the UK and the USA. Our list includes only these companies from which we have purchased seeds; there are plenty of others that also sell worthwhile varieties.**

UK

**D.T. Brown & Co.**
Station Road
Poulton-le-Fylde
Lancashire FY6 7HX
tel 01253 882 371
fax 01253 890 923

**Chiltern Seeds**
Bortree Stile
Ulverston
Cumbria LA12 7PB
tel 01229 581 137
fax 01229 584 549

**Mr Fothergill's Seeds**
Kentford
Newmarket
Suffolk CB8 7QB
tel 01638 552 512
fax 01638 751 624

**Future Foods**
P.O. Box 1564
Wedmore
Somerset BS28 4DP

**E.W. King Co. & Suffolk Herbs**
Monks Farm
Kelvedon
Colchester
Essex CO5 9PG
tel 01376 572 456
fax 01376 571 189

**Edwin Tucker & Sons**
Brewery Meadow
Stonepark
Ashburton
Newton Abbot
Devon TQ13 7DG
tel 01364 652 403
fax 01364 654 300

**The Organic Gardening Catalogue**
River Dene Estate
Molesey Road
Hersham
Surrey KT12 4RG
tel 01932 253 666
fax 01932 252 707

**Poyntzfield Herb Nursery**
Black Isle, By Dingwall
Ross & Cromarty
Scotland IV7 8LX
tel/fax 01381 610 352

**Seeds-By-Size**
45 Crouchfield
Boxmoor
Hemel Hempstead
Hertfordshire HP1 1PA
tel 01442 251 458

**Simpson's Seeds**
27 Meadowbrook
Old Oxted
Surrey RH8 9LT
tel/fax 01883 715 242

**Suttons**
Hele Road
Torquay
Devon TQ2 7QJ
tel 01803 614 614
fax 01803 615 747

**Thompson & Morgan**
Poplar Lane
Ipswich
Suffolk IP8 3BU
tel 01473 688 588
fax 01473 680 199

USA

**Guerney's Seed & Nursery Co.**
110 Capital Street
Yankton
South Dakota 57079
tel 605 665-1930
fax 605 665-9718

**Harris Seeds**
60 Saginaw Drive
P.O. Box 22960
Rochester
New York 14692-2960
tel 716 442-0410
fax 716 442-9386

**Johnny's Selected Seeds**
1 Foss Hill Road
RR1 Box 2580
Albion
Maine 04910-9731
tel 207 437-4395
fax 207 437-2165

**Nichols Garden Nursery**
1190 North Pacific Highway
Albany
Oregon 97321-4580
tel 541 928-9280
fax 541 967-8406

**Park Seed**
1 Parkton Avenue
Greenwood
South Carolina 29647-0001
tel 864 223-7333
fax 864 941-4206

**Pinetree Garden Seeds**
Box 300
New Gloucester
Maine 04260
tel 207 926-3400
fax 888 527-3337

**Seeds Blum**
Idaho City Stage
Boise
Idaho 83706
tel 208 342-0858
fax 208 338-5658

**Shepard's Garden Seeds**
30 Irene Street
Torrington
Connecticut 06790
tel 860 482-3638
fax 860 482-0532

**Stokes Seeds Inc.**
P.O. Box 548
Buffalo
New York 14240-0548
tel 716 695-6980
fax 888 834-3334

**Sunrise Enterprises**
P.O. Box 1960
Chesterfield
Virginia 23832
tel 804 796-5796
fax 804 796-6735

**Totally Tomatoes**
P.O. Box 10
571 Whaley Pond Road
Graniteville
South Carolina 29829-0010
tel 803 663-9771
fax 888 477-7333

**Willhite Seed Inc.**
P.O. Box 23
Poolville
Texas 76076
tel 800 828-1840
fax 817 599-5843

# BIBLIOGRAPHY

Andeweg, J.M. and De Bruyn, J.W. 1959. Breeding of Non-bitter Cucumbers. *Euphytica.* 8:13-20.
Andrews, J. *Peppers: The Domesticated Capsicums.* University of Texas Press, Austin, Texas, 1990.
Atherton, J.G. and Rudich, J. (eds.) *The Tomato Crop.* Chapman and Hall, London, 1986.
Bareham, L. *In Praise of the Potato.* Michael Joseph, London, 1989.
Bassett, M.J. (ed.) *Breeding Vegetable Crops.* AVI Publishing Co., Westport, Conn., 1986.
Belleme, J. and Belleme, J. *Culinary Treasures of Japan.* Avery, New York, 1992.
Bissell, F. *Sainsbury's Book of Food.* J Sainsbury, London, 1989.
Bleasdale, J.K.A., Salter, P. J. and others. *The Complete Know and Grow Vegetables.* Oxford University Press, Oxford, 1991.
Blythman, J. *The Food We Eat.* Michael Joseph, London, 1996.
Bose, T.K. and Som, M.G. (eds) *Vegetable Crops in India.* Naya Prokash, Calcutta, 1986.
Bosland, P.W., Bailey, A.L. and Iglesias-Olivas, J.*Capsicum Pepper Varieties and Classification.* Circular 530, Cooperative Extension Service, College of Agriculture and Home Economics, New Mexico State University, Las Cruces, New Mexico, 1996.
Brewster, J.L. *Onions and Other Vegetable Alliums.* CAB International, Wallingford, Oxford, 1994.
Brickell, C. (ed-in-chief). *The Royal Horticultural Society Encyclopedia of Gardening.* Dorling Kindersley, London, 1993.
Bown, D. *The Royal Horticultural Society Encyclopedia of Herbs and Their Uses.* Dorling Kindersley, London, 1995.

Chowings, B. 1997. Adding Bite to Salads. *The Garden*. 122:416-417.
Cost, B. *Foods from the Far East*. Century, London, 1990.
Coultate, T. and Davies, J. *Food: the Definitive Guide*. Royal Society of Chemistry, Cambridge, 1994.
Council of Scientific and Industrial Research. *Wealth of India*. Publication and Information Directorate, New Delhi, 1956-1992.
Department of Health. *Dietary Reference Values for Food Energy and Nutrients for the United Kingdom*. HMSO, London, 1991.
Department of Health. *Nutritional Aspects of the Development of Cancer*. HMSO, London, 1998.
DeBaggio, T. and Belsinger, S. *Basil: An Herb Lover's Guide*. Interweave Press, Loveland, Colorado, 1996.
Devi, Y. *The Art of Indian Vegetarian Cookery*. Leopard, London, 1995.
DeWitt, D. and Bosland, P.W. *Peppers of the World*. Ten Speed Press, Berkeley, California, 1996.
Dunbar, R. 1998. Bizarre Brassicas. *The Garden*. 123:191-193.
Gaman, P.M. and Sherrington, K.B. *The Science of Food*. Pergamon, Oxford, 1981.
Gray, A.R. 1989. Taxonomy and Evolution of Broccolis and Cauliflowers. *Baileya*. 23:28-46.
Greene, B. *Greene on Greens*. Equation, Wellingborough, Northamptonshire, 1987.
Grigson, J. *Jane Grigson's Vegetable Book*. Penguin, London, 1980.
Haroutunian, A. der *Classic Vegetable Cookery*. Ebury Press, London, 1985.
Harris, C.C. and Howells, M. *Modern Ways of Growing and Cooking Vegetables*. Garden Book Club, London, 1972.
Heal, C. and Allsop, M. *Queer Gear: How to Buy and Cook Exotic Fruits and Vegetables*. Century Hutchinson, London, 1986.
Henry Doubleday Research Association. *The Fruit and Veg Finder*. Ryton-on-Dunsmore, Coventry, 1995.
Holland, B., Unwin, I.D., Buss, D.H. *Vegetables, Herbs and Spices: The Fifth Supplement to McCance & Widdowson's The Composition of Foods* (4th Edition). Royal Society of Chemistry, Ministry of Agriculture, Fisheries and Food, Cambridge and London, 1991.
Hom, K. *Asian Ingredients*. Ten Speed Press, Berkeley, California, 1996.
Jaffrey, M. *A Taste of India*. Pavilion, London, 1985.
Jaffrey, M. *Eastern Vegetarian Cooking*. Jonathan Cape, London, 1983.
Jones, H.A. and Mann, L.K. *Onions and their Allies*. Leonard Hill, London, 1963.
Kalloo, G. and Bergh, B.O. *Genetic Improvement of Vegetable Crops*. Pergamon Press, Oxford, 1993.
Kennedy, D. *The Art of Mexican Cooking*. Bantam, New York, USA, 1989.
Kraig, B. and Nieto, D. *Cuisines of Hidden Mexico: a Culinary Journey to Guerrero and Michoacán*. John Wiley & Sons, New York, 1995.
La Place, V. *Verdura*. Macmillan, London, 1994.
Larkcom, J. *The Salad Garden*. Frances Lincoln, London, 1984.
Larkcom, J. *Oriental Vegetables*. John Murray, London, 1991.
Librairie Larousse. *Larousse Gastronomique*. Hamlyn, London, 1988.
Lo, K. (ed.) *The Complete Encyclopedia of Chinese Cooking*. St Michael, Octopus Books, London, 1981.
Macrae, R., Robinson, R. and Sadler, M. (eds.) *Encyclopaedia of Food Science, Food Technology and Nutrition*. Academic Press, London, 1993.
McClure, S. *The Harvest Gardener*. Garden Way, Pownal, Vermont, 1993.
McGee, H. *On Food and Cooking: The Science and Lore of the Kitchen*. Allen & Unwin, Hemel Hempstead, Hertfordshire, 1986.
Miller, M. *The Great Chile Book*. Ten Speed Press, Berkeley, California, 1991.
Ministry of Agriculture, Fisheries and Food. *Manual of Nutrition*. HMSO, London, 1995.
Monselise, S. P. (ed.) *CRC Handbook of Fruit Set and Development*. CRC Press, Inc. Boca Raton, Florida, 1986.
National Consumer Council. *Your Food: Whose Choice?* HMSO, London, 1992.
Nayar N.M. and More, T.A. (eds) *Cucurbits*. Science Publishers, Inc., Enfield, New Hampshire, 1998.
Nothmann, J., Rylski, I. and Spigelman, M. 1976. Color and Variations in Color Intensity of Fruit of Eggplant Cultivars. *Scientia Horticulturae*. 4:191-197.
Ory, R.L. *Grandma Called it Roughage*. American Chemical Society, Washington DC, 1991.
Owen, S. *Healthy Thai Cooking*. Frances Lincoln, London, 1997.
Picha, D.H. and Hall, C.B. 1982. Effect of Potassium Fertilization and Season on Fresh Market Tomato Quality Characters. *HortScience*. 17:634-635.
Platt, K. *The Seed Search* (Second Edition). Karen Platt, Sheffield, 1997.
Price, K.R., DuPont, M.S., Shepherd, R., Chan, H.W-S. and Fenwick, G.R. 1990. Relationship between the Chemical and Sensory Properties of Exotic Salad Crops – Coloured Lettuce (*Lactuca sativa*) and Chicory (*Cichorium intybus*). *Journal of the Science of Food and Agriculture*. 53:185-192.
Rick, C.M. 1978. The Tomato. *Scientific American*. 239:67-77.
Robinson, R.W. and Decker-Walters, D.S. *Cucurbits*. CAB International, Wallingford, Oxford, 1997.
Roden, C. *A New Book of Middle Eastern Food*. Penguin, London, 1986.
Rohde, E.S. *Uncommon Vegetables*. Country Life, London, 1943.
Rohde, E.S. *Vegetable Cultivation and Cookery*. Medici Society, London, 1938.
Ross, R.L.S. *Beyond Bok Choy: a Guide to Asian Vegetables*. Artisan, New York, 1996.
Rozin, E. *Blue Corn and Chocolate*. Ebury Press, London, 1992.
Rubatzky, V.E. and Yamaguchi, M. *World Vegetables: Principles, Production, and Nutritive Values* (Second Edition). Chapman & Hall, New York, 1997.
Ryder, E.J., De Vos, N.E. and Bari, M.A. 1983. The Globe Artichoke (*Cynara scolymus* L.). *HortScience*. 18:646-653.
Seymour, G.B., Taylor, J.E. and Tucker, G.A.(eds.) *Biochemistry of Fruit Ripening*. Chapman & Hall, London, 1993.
Smith, D. *Food Watch*. HarperCollins, London, 1994.
So, Y. *Yan-Kit's Classic Chinese Cookbook*. Dorling Kindersley, London, 1984.
Stone, S. and Stone, M. *The Essential Root Vegetable Cookbook*. Potter, New York, 1991.
Suzuki, M. and Cutcliffe, J.A. 1981. Sugars and Eating Quality of Rutabagas. *Canadian Journal of Plant Science*. 61:167-169.
Tanis, D. *Corn*. Collins, San Francisco, 1995.
Tolonen, M. *Vitamins and Minerals in Health and Nutrition*. Ellis Horwood, Chichester, West Sussex, 1990.
Vaughan, J.G. and Geissler, C.A. *The New Oxford Book of Food Plants*. Oxford University Press, Oxford, 1997.
Veris Research Information Service. *Carotenoids Fact Book*. La Grange, Illinois, 1996.
Vilmorin-Andrieux, M.M. *The Vegetable Garden*. 1885 (Facsimile Edition). Ten Speed Press, Berkeley, California.
Virani Food Products. *The Flavours of Gujarat*. Wellingborough, Northamptonshire, 1991.
Waters, A. *Chez Panisse Vegetables*. HarperCollins, New York, 1996.
Whealy, K. *Garden Seed Inventory* (4th Edition). Seed Saver Publications, Decorah, Iowa, USA, 1995.
Wien, H.C. (ed.) *The Physiology of Vegetable Crops*. CAB International, Wallingford, Oxford, 1997.
Willan, A. *Reader's Digest Complete Guide to Cookery*. Dorling Kindersley, London, 1989.
Wilson, A. *The Story of the Potato Through Illustrated Varieties*. Alan Wilson, 1993.
Wine and Food Society, The. *A Concise Encyclopedia of Gastronomy, Section III, Vegetables*. London, 1941.
Woolfe, J.A. *The Potato in the Human Diet*. Cambridge University Press, Cambridge, 1987.
World Cancer Research Fund/American Institute for Cancer Research. *Food, Nutrition and the Prevention of Cancer: a Global Perspective*. American Institute for Cancer Research, Washington DC, USA, 1997.
World Health Organization *Diet, Nutrition and the Prevention of Chronic Diseases*. Geneva, 1990.

# INDEX

Page numbers in *italic* refer to illustrations.

# ACKNOWLEDGMENTS

We are deeply indebted to Frances Lincoln for agreeing to publish this book, and to Jo Christian, Lewis Esson, Louise Kirby, Sally Cracknell, Kate Phillips, Patricia Going, Jo Grey and Tom Armstrong for their expertise and encouragement. Thanks and appreciation also to James Merrell for the beautiful photography.

Christine McFadden would particularly like to thank Ed McFadden for his good humour and support; Natalie Jiggins for deciphering and typing the recipes; Katy Balfry for her assistance in preparing dishes for photography; Greg Wallace and Charlie Hicks of the George Allen Company for providing out-of-season vegetables; and Sri Owen for advice on Thai cooking.

Michael Michaud would like to thank the following for providing vegetables: Hugh and Patsy Chapman; Stewart Clarke, West Deane Gardens; Georgina Connaughton; Andrew and Sarah Cross; Simon Eastwood; Toos Jeuken; Keith Martin; and Ian Nelson. Thanks are also extended to the 'staff' at Sea Spring Farm: Sean Skinner, Paula Knight and Rowan Waldron. Much-appreciated technical help was provided by the following: Dr. Peter Crisp, Crisp Innovar, Ltd.; Mike Day, National Institute of Agricultural Botany; Dr. Jonathan Davey, Scottish Agricultural Science Agency; Dr Frances Gawthrop, James Hatherill and Steve Winterbottom, A.L. Tozer Ltd.; Joy Larkcom; Dr Richard N. Lester, Birmingham University; Dr Tony Lord; and Barbara Segal. Much-needed books were tracked down by Jon and Sue Atkins, Summerfield Books; Keith Crotz, American Botanist Booksellers; David and Diana Leake, The Bookshop. Thanks to my children, Ben and Martha, for their patience and understanding; Susan Anderson, for encouragement when it was most needed; and Arthur Pearse, for sharing many gardening experiences.

**Editor and Project Manager**: Lewis Esson
**Horticultural Consultant**: Tony Lord
**Indexer**: Antonia Johnson
**Production**: Liz Stewart
**Commissioning Editor**: Jo Christian
**Art Editors**: Louise Kirby & Sally Cracknell
**Picture Editor**: Anne Fraser
**Editorial Director**: Erica Hunningher
**Art Director**: Caroline Hillier